LEFT AND RIGHT OF THE BOOM

The Art of Leadership

Before, During, and After Crisis

Praise for *Left and Right of the Boom*

"Anywhere, at any time, in any organization, things can blow up. At that moment, everyone needs a leader who is steadfast, calm, and engaging. Rob Campbell does an amazing job drawing upon his 27 years of military leadership and life experiences as a business owner to present practical lessons and riveting stories from the battlefield in *Left and Right of the Boom*. You will learn time-tested principles of how to lead effectively before, during, and after a crisis. This book is a must-read for any business owner or executive leader."

Dave McAuley
Founder and Leadership Coach at Summit Leadership

"Rob Campbell's third book, *Left and Right of the Boom*, is another remarkable capture of battlefield and military lessons for those who lead in crisis. Leaders and organizations would be well-served by reading this book and heeding its lessons ... because a crisis is coming."

General James T. Hill
United States Army (Retired)

"I feel the pressure of crisis each day as a business leader. I'm thankful I have Rob as a coach to help discover tools to overcome threats and the confidence to thrive. Now, with this timely and impactful book, you too can benefit from Rob's experience and lessons. Lesson #1: Accept that a crisis is coming. *Don't* accept that you are not ready for it. Get Left and Right of the Boom!"

Georgette Heller
Marketing and Communications Leader
DRTV Senior Director, TRANZACT

"Every page in *Left and Right of Boom* is as explosive as its title. Rob Campbell superbly captures the essence of leadership, and why leadership matters during crisis. Each page not only captivates you but gives detailed insight as to why preparing for the inevitable is so important before, during, and after events occur. This book is a valuable resource for any business — a treasure trove of knowledge and experience that has been battle-tested, in combat and in today's business."

Colonel Adam L. Rocke
United States Army (Retired)
President, Client Relations at Katie's Way and Katie's Way Plus

"Do you ever wonder why the military works so hard to create great leaders? Because unlike the business and athletic worlds — where failed leaders are still paid vast amounts of money and organizations might still function relatively well — having poor leaders in the military means more people die. In his new book, Colonel Rob Campbell captures the leadership behaviors that *every* coach/leader/manager needs to follow. He vulnerably shares his own leadership achievements, failures, and learnings, then articulates in actionable ways how to successfully apply a healthy leadership model in our civilian world. Thank you, PFC Patrick Devoe, for your service. Thank you, Colonel Rob Campbell, for using that 'Boom' to create some good for all mankind."

Bill Eckstrom
Founder and Chairman, The Ecsell Companies

"I've always believed that individual competence consists of knowledge and skill combined with discipline and a positive attitude. Individual competence leads to individual confidence, unit competence, and winning teamwork — and ultimately mission success. And in times of crisis, competence and confidence must not be in short supply. Rob Campbell's book begins with a crucible of crisis — the loss of a soldier. It ends with a call to each of us to get 'left and right' of our own 'booms.' The pages in between are packed with Rob's journey — gaining competence and confidence — and demonstrates how you can too. This book, written by a truly competent, authentic leader, is a must-read for every leader."

Lawson W. Magruder, III
Lieutenant General, U.S. Army (Retired)
Author of A Soldier's Journey Living His Why: Inspired by a Little Green Book

"Rob Campbell knows crisis first-hand. As a former senior Army leader and now business owner, he has weathered many difficult situations and learned the value of preparing for, navigating, and learning from crisis. As Rob explains, it starts with investing in people and preparing the organization and culture to be resilient and ready to step up when challenges arise. I witnessed Rob in action, serving as one of his junior commanders in a combat zone, and his cool, collected, steady leadership anchored the larger organization in times of crisis. This book is a must-read for leaders, entrepreneurs, and aspiring leaders, as it will better equip you to face 'the Boom' that is an inevitable part of the leadership journey!"

Christian White
President & Co-founder, risk3sixty, LLC

"I began my Army career with Rob 34 years ago and have been learning from him ever since. His book underscores that life is a series of crises — often unexpected and potentially devastating. Rob provides crucial, hard-won insights to equip and inspire you and your team to prepare for, adapt during, and learn from each crisis in an iterative cycle that builds capacity and resilience. I recommend reading and talking about this book with your team. You will be better for it."

Colonel Tony Burgess
United States Army (Retired)
Co-Author, Taking the Guidon: Exceptional Leadership at the Company Level
Founder, The Cornwall Leadership Institute

"Colonel Rob Campbell has spent a professional lifetime thinking about, preparing for, responding to, and learning from crisis. If you can commit yourself to the chapters of this remarkable book, every crisis you encounter - for the rest of your life - will feel less daunting, less overwhelming, less devastating, and less impactful."

Mike Barger, EdD
Professor, University of Michigan's Ross School of Business
Author of High-Stakes Leadership in Turbulent Times: Why Stakeholders Are Your Greatest Assets ... in Good Time and Bad
JetBlue Airways Co-founder
Former Navy TOPGUN Chief Instructor

LEFT AND RIGHT OF THE BOOM

The Art of Leadership
Before, During, and After Crisis

Colonel Rob Campbell
UNITED STATES ARMY, RETIRED

WITH A FOREWORD BY
Mike Barger, Co-Founder of JetBlue Airways

BENTMOUTH PRESS

Left and Right of the Boom: The Art of Leadership
— Before, During, and After Crisis

Published by Bentmouth Press (Johnson City, TN).

Left and Right of the Boom is Book 3 in the *Battlefield to the Boardroom Series* by Colonel Rob Campbell.

Editing by:
Kate Colbert

Cover design and typesetting by:
George Stevens

First edition, July 2024

Paperback ISBN: 979-8-9907126-0-7
Hardcover ISBN: 979-8-9907126-1-4
Library of Congress Control Number: 2024909549

Created in the United States of America

Dedication

This book is dedicated to Private First Class Patrick Devoe, II

Killed in Action
March 8, 2009
Operation Enduring Freedom, Afghanistan

And to all the fallen of our wars following 9/11 ...

May they be enshrined on the monuments and in the memories of this nation forevermore. And may we be guided — every day — by the lessons of loyalty, love, and leadership they taught us through their service and their lives.

Also by Colonel Rob Campbell

It's Personal, Not Personnel: Leadership Lessons for the Battlefield and the Boardroom

At Ease: Enjoying the Freedom You Fought For — A Soldier's Story and Perspectives on the Journey to an Encore Life and Career

Contents

"All people — all lives — are either in a crisis, coming out of a crisis, or headed for a crisis."

Andy Andrews

Foreword

By Mike Barger, Co-founder, JetBlue Airways

At 8:46 a.m. on September 11, 2001, American Airlines Flight 11 ripped through the North Tower of the World Trade Center in New York City. As the Executive Director of JetBlue Airways' Emergency Command Center (ECC), I was notified almost immediately that an unknown aircraft had crashed in Lower Manhattan. *Could it be one of ours?*

Boom!

At that moment of impact, JetBlue Airways was 20 months old. We were "New York's Hometown Airline," with our flying base at JFK Airport and our headquarters in Queens, NY, about 10 miles east of Lower NYC. We were operating a fleet of 17 Airbus 320 aircraft and had placed orders to add an additional plane every five weeks for the next 10 years. By any measure, we had planned the most ambitious start-up in commercial aviation history.

That fateful morning, now recalled simply as "9/11" around the world, I was welcoming a group of new hires in our Queens headquarters. The back windows of our seventh-floor classroom faced west, and the southern New York City skyline could be easily seen in the distance on this beautiful, blue-sky morning. A member of my ECC team came dashing into the room, and I immediately knew something had gone terribly wrong. *Why was someone from the emergency command center bolting into an onboarding session, out of breath and with a look of horror on her face?* She informed me of the crash at the Word Trade Center (WTC) and pointed toward the back of the room. Through the classroom windows, I could see a column of black smoke starting to rise from the southern cluster of buildings.

I raced down two flights of stairs to our ECC, where we had a bank of televisions that were installed to help us follow media coverage from the major networks whenever JetBlue needed to activate its Command Center. Images of the WTC's smoldering North Tower filled each of the screens. Other senior JetBlue leaders arrived as I did, and as we clamored to determine if the aircraft in question was ours, we watched another aircraft — United Airlines Flight 175 — crash into the South Tower. It was 9:03 a.m. *Are we at war?*

Boom!

A flurry of activity followed at JetBlue headquarters — but not a confused, random assortment of actions. Rather, company leaders set into motion a documented, rehearsed, and well-understood set of practices that had been crafted well before JetBlue had taken delivery of its first aircraft. JetBlue formally activated its Command Center, alerting ECC representatives to report immediately to their emergency-response stations. We also began to send out communications to all employees ("crewmembers," we called them) across the airline, to ensure everyone knew what was happening. The practice of radical transparency and over-communication were strongly held beliefs across the leadership team at JetBlue and they were about to serve us well. We believed that exceptional execution, under any circumstances, would require everyone at the company to be situationally aware and prepared to assist if, when, and where necessary.

Once it was determined that neither WTC impact involved a JetBlue aircraft, priorities quickly shifted to the safety of other JetBlue planes, passengers, and crew who could possibly be facing extreme, unprecedented danger. JetBlue had nine flights in the air — far fewer than any major carrier, but certainly no less important. We began communicating with all of them, doing our best to share what we knew at the time, which was very little, other than what we could see on our ECC TVs.

At 9:42 a.m., less than an hour after the first plane struck the Towers, the U.S. Federal Aviation Administration (FAA) ordered all airborne civilian aircraft (there were about 4,500 of them!) to land as soon as possible. They were ordered to land not at the nearest airport where there were other company employees, but at the nearest airport where a safe landing could be accomplished — irrespective of the support that would or would not be there to assist in parking or deplaning the aircraft or transporting, lodging, or caring for the people on those flights. The National Airspace System was closing. JetBlue leaders, along with those of every other commercial airline, were faced with many defenseless aircraft in the air; the horrific news of several confirmed simultaneous hijackings; and a best-case scenario of aircraft, customers, and crew being stranded at unexpected and unsupported locations across the country. We had no idea what precipitated the hijackings — why and how they happened or whether there was a common thread among them — nor did we have any sense of what the future might hold. But there we were, dealing with an unimaginable crisis. At 11:06 a.m., the FAA issued Advisory 036, officially shutting down the skies.

Boom!

In the three days that followed the first-ever unplanned total closure of U.S. airspace, the leaders assigned to JetBlue's ECC unwaveringly dedicated themselves to supporting people, planes, families, and countless members of the company's extended network of stakeholders. This was the first time in JetBlue's history that it was seriously challenged to *walk the talk* — to live up to the commitment declared in the company's vision statement of "Bringing humanity back to air travel" and its core values of "Safety, Caring, Integrity, Fun, and Passion." As I look back on what the ECC team was able to accomplish as a unit — and the countless acts of kindness, caring, empathy, courage, and individual heroism demonstrated by members of JetBlue's extended community — I am awestruck by the potential of humankind when times are at their worst. But I'm also reminded that we perform best under pressure when given a place to start — a foundation upon which we can effectively launch a response to the most challenging circumstances that a volatile, uncertain, complex, and ambiguous world can throw at us.

We perform best under pressure when given a place to start — a foundation upon which we can effectively launch a response to the most challenging circumstances that a volatile, uncertain, complex, and ambiguous world can throw at us.

Rob Campbell's book, which you have wisely laid your hands on, examines that foundation and teaches you how to build that "place to start" so you're not caught totally unawares when crisis strikes your organization.

In 2001, with JetBlue less than two years old (at a time when most start-ups are so focused on growth that they can't even imagine disruption), we were ready for crisis. Some of that foundation was inherent in the industry — commercial aviation is a highly regulated, high-stakes, life-and-death venture. But much of our readiness was owing to our culture, which we created deliberately and with heart. JetBlue was founded to be much more than just another airline. We set out to build a great company that people would want to be a part of, that stakeholders would want to do business with, and that was prepared as much as it could be for the inevitable series of disruptions and crises that would surely come its way. Our vision, values, and leadership principles collectively served as our North Star, guiding our actions and decision-making in good times and bad. But without thoughtful preparation and a company-wide commitment to readiness for the worst of times, JetBlue would never have been able to demonstrate such remarkable resilience then, in 2001, or during the past 25 years leading us to today. It is this commitment, I believe, that has truly set JetBlue apart. Time will tell if the JetBlue business model allows the company to enjoy success in the decades to come, but in terms of being prepared for the inevitable crises that it will most certainly have to face, there is not a more capable enterprise on the planet.

I share this story as an introduction to the lessons you are about to explore in this emotional and thoughtful book by my friend, Rob Campbell. Each "boom" in this foreword was provided to illustrate a critical moment for JetBlue leaders — a moment that would require:

- Calm, clear, and precise thoughts and actions
- Exceptional communication with a wide assortment of stakeholders
- Great courage to face each challenge head-on
- Exceptional confidence that obstacles, no matter how daunting, can be overcome, and
- A willingness to creatively find the very best of many possible solutions to move the team, the unit, and the enterprise forward.

In the pages that follow, you will become very familiar with these concepts.

Whatever your motivation for picking up a book entitled *Left and Right of the Boom: The Art of Leadership Before, During, and After a Crisis,* I imagine that you are either curious about what you have seen from crisis leaders in the past or you are wondering what you can do to prepare yourself and your team for the challenges today's business environment has in store for you. Perhaps both. I applaud your curiosity and can validate that it is well-placed. What you are now holding in your hands (or viewing on a digital screen) is a treasure trove of remarkably practical and incredibly valuable guidance that will help you prepare for, deal with, and learn from your next crisis. I'm very sorry to say — and this is certainly not breaking news — that your next crisis will not be your *last.* But if you can commit yourself to the chapters that follow, every crisis you encounter — for the rest of your life — will feel less daunting, less overwhelming, less devastating, and less impactful.

Colonel Rob Campbell has spent a professional lifetime thinking about, preparing for, responding to, and learning from crisis. As I reflect on my own professional career, I feel as though I have done the same. We both have a great deal of military experience — Rob's leading massive ground forces in the U.S. Army and mine flying fighters in the U.S. Navy. Although our warfighting paths were different, we both learned that there are a core set of proven, battle-tested ways to effectively lead in the incredibly challenging operational environments that our service members are facing today. In our post-military careers, we have both learned that these leadership practices and principles are just as effective in the boardroom as they are on the battlefield. Rob has shared an exceptional collection of these, along with clear explanations and compelling examples, in this book.

If you can commit yourself to the chapters that follow, every crisis you encounter — for the rest of your life — will feel less daunting, less overwhelming, less devastating, and less impactful.

We have also learned over the years that organizational crises are not simply functions of one-in-a-million, "Black Swan" events, such as those I described in my 9/11 story or that we all experienced during the COVID-19 pandemic. No, significant disruptions and crises are the new normal. Why? Because the pace of change in all aspects of our world has never been faster. And this pace of change will continue to accelerate, driven by many factors. To name a few, leaders will have to deal with:

- Constantly evolving and improving technology, from smaller, faster, more powerful microprocessors to generative artificial intelligence
- The availability of information and the constant challenge of sorting fact from fiction
- The ubiquity of social media and the speed at which news — real and manufactured — can reach millions in an instant, and
- The rising tensions between groups of people who view the world differently.

Each of these will require strong leadership to navigate — and each will have a significant impact on your own, your team's, and your organization's ability to perform.

For an enterprise to be successful going forward, senior-most organizational leaders must give serious thought to how they will prepare everyone — themselves, leaders throughout the organization, and individual contributors — across the company not just for operational excellence (which, on its own, requires great effort), but also for a broad spectrum of potential circumstances and scenarios, obstacles and challenges, disruptions and crises. It won't be easy. It will feel daunting. But in this book, Rob provides a great place to start.

In the pages of *Left and Right of the Boom*, you have discovered the right resource to help you prepare for the worst, while building a set of skills and capabilities that will significantly enhance your ability to perform at your best — in good times and bad.

Mike Barger, EdD
Professor, University of Michigan's Ross School of Business
Author of High-Stakes Leadership in Turbulent Times: Why Stakeholders Are Your Greatest Assets ... in Good Time and Bad
JetBlue Airways Co-founder
Former Navy TOPGUN Chief Instructor

Introduction

On the morning of March 8, 2009, I awoke after a restless night's sleep on a remote base near the border of Pakistan. I was only two days away from assuming control of a battlefield — roughly the size of Rhode Island — and a combat task force of more than 700 men and women (soldiers and civilians) serving during Operation Enduring Freedom, the surge of forces in Afghanistan. I was a United States Army Lieutenant Colonel and the Commander of 1st Squadron, 40th Cavalry Regiment, Airborne. In my tenure as commander of this parachute unit, I would add another chapter of combat to the unit's storied history dating back to World War II. It would be the pinnacle of my military career to date, leading at that level in a combat zone.

March 8th was a beautiful morning. The cedar trees were budding, and dwarf cherry shrubs were ablaze with delicate pink blossoms filling the air with their subtle fragrance — a perfect contrast against a landscape deeply scarred by decades of war. All around me were reminders that I was far from home, in a combat zone: hints of smoke from ovens cooking naan bread and the noise of armored vehicles, soldiers securing combat gear, and sergeants barking orders. The wind was calm and the temperature perfect — eerily like the weather in New York City and Washington, DC, and in the skies over Shanksville, Pennsylvania, on September 11, 2001. Spring had arrived in the Hindu Kush, the mountain range that dominated Eastern Afghanistan. So, too, had the summer fighting season, so named for its increase in violent attacks by the Taliban. That anxious feeling pulling at my gut — that feeling that troubled my sleep last night — returned to visit me. It was an all-too-familiar experience in my military career and a previous year preparing for war.

That morning, I found myself in a makeshift conference room with my fellow commanders, taking a briefing from the unit we would be replacing. Minutes into the opening briefing, a young officer dashed into the room and whispered something in the ear of the senior commander. The senior commander — a tough, demanding, West Point officer with combat service in Panama and Iraq and with whom I had served previously in the famed 82nd Airborne Division — quickly turned to me and directed, "Rob, you need to head to the TOC." A TOC is the Tactical Operations Center, the hub of all activity in a combat zone. The look of horror on his face was a countenance I had seen before. I was already a combat veteran with service in Iraq. This was not good news. My blood ran cold, and I stiffened as I heard his words. I jumped up from my chair and rushed down the narrow hallway — adorned with pictures of fallen warriors from the previous year — and I entered the operations center.

The TOC was awash with activity — people sharing and shouting updates, digital maps on large video monitors displaying current activity, and pensive radio chatter delivering an extremity of clamor and bustle, creating the kind of tension one feels in their chest when under threat. I felt the pressure increase, as if I was being squeezed between two walls, yet strangely I could not hear any of it. As if being able to mute it, I focused instead on the watch officer — the senior officer in the operations center, who was orchestrating reports, systems, and people. I knew he could help me cut through the chaos and tell me what was happening. The watch officer took notice of my arrival and gave me an update on what had just happened. A military Humvee had struck an improvised explosive device (IED) and a soldier was killed.

My soldier.

Private First Class Patrick Devoe from Auburn, New York.

"Devoe," as we called him, was not a typical young soldier. He entered the Army at an older age of 27. He was the daddy of his squad, up early each morning, making coffee and tending to his buddies. We would rely on men like him to do the impossible in the year ahead. In an instant, he was taken from this earth, his family, and his unit. For a new Army recruit, 27 is "older." But suddenly, as a fallen soldier, he felt so very young. His death hit me like a truck. I had prepared for moments like this in training, but this reality — an actual loss of one of my men — left me momentarily paralyzed. The news of Devoe's death would spread like wildfire and traumatize leaders and soldiers across my command on the front end of a 12-month combat tour. My organization was suddenly in the thick of a truly high-stakes moment. It felt like a crisis, but I was too focused to define it just yet. It was surely a tragedy. And it would require the best of my leadership.

Boom! On the Battlefield and in the Boardroom

I don't expect that everyone reading this book has experience as a combat veteran. In fact, I expect that most of you don't. But what I *do* know to be true is that all of us who are privileged to lead will eventually find ourselves in the middle of a crisis ("the boom"), dealing with the fallout after crisis ("right of the boom"), and eventually planning and preparing for the next crisis ("left of the boom"). The truth of the matter is that anywhere and in any kind of organization — military or civilian — things can blow up. The secret to avoiding such explosions (literal and figurative), reducing them in number or severity, and reacting to them as best as possible comes down to leaders who know how to think constantly about what to *do* and how to *be* when they are "left and right of the boom" (i.e., before and after the crisis).

The truth of the matter is that anywhere and in any kind of organization — military or civilian — things can blow up.

I've spent a lot of years thinking about how military units behave when they're left and right of the boom, and I've long thought that what I know about this topic could help leaders in any industry. I've been wanting to write this book for you for quite some time. And then, *BOOM*! A global pandemic. It was March 2020, and the entire world was thrust into a crisis unlike anything we'd experienced in our lives. In some ways, this was familiar to me. My body and mind knew how to jump into "response mode" — assessing the unthinkable situation and making quick plans for what to do next. Like everyone else, I stayed busy in that "right of the boom" space for several months, which became a few years. Then I sat down to write this book because I thought the time had come — a time when I could steel myself against the difficult emotions of reliving some of my most traumatic crises to teach you some lessons about how to handle *your* next awful moment when things go very horribly, even tragically, wrong.

My goal with this book is to help you harden your organization and fortify your leadership mettle for imminent crisis. Because the truth is stark: it's not *if* crisis will befall your organization, it's *when*. Take your pick — pandemic, inflation, sudden loss of a critical employee, or a seemingly overnight change in technology rendering your product or service irrelevant, the next challenge for your company is right around the corner. Preparing for or engaging in combat, one could argue, is easy in terms of crisis expectation, preparation, and response. The enemy will engage. Soldiers will be injured or killed. Because of this, military leaders and their teams "lean forward in the foxhole" — alert and ready — scanning the horizon for threats. A business, in contrast, will live in environments of relative peace and order — where systems work, people perform, and customers respond favorably. The stories and concepts in this book will help you "lean forward in your own foxhole."

Muscle Memory for Crisis Leadership

In times of crisis, it's imperative that we turn not to emotions and erratic thought but to reason, principles, values, and training. I implore you to pause, if only momentarily, when crisis strikes and remind yourself that you've "been here before" in a training scenario or a previous crisis or operational disruption. Maybe in that moment of pause, you'll even remember what I'm about to offer in this book, *Left and Right of the Boom*. I wrote this book for you so it might serve as your personal training for every "boom!" that awaits you. The power of reflection — of coming back to the muscle memory of crisis response — has served me well when things have blown up in my life and career. In countless scenarios played out on training grounds, in classrooms, and on battlefields, I was guided indeed by reason, principles, and my core beliefs, moral code, and values, but mostly by my training. Practice may or may not "make perfect," but having been there — in body and spirit — allows us to start from a place of experience when crisis strikes.

In times of crisis, it's imperative that we turn not to emotions and erratic thought but to reason, principles, values, and training.

When I look back at my 27-year career in the U.S. Army and my subsequent career as a leadership consultant and small-business owner, there's no shortage of examples of me drawing from my past to handle a quagmire in the present. Notably, this happened — for the most part — subconsciously; my memory would bring back concepts, lessons learned, and successes and failures that returned to me at the most appropriate times. I consider myself a successful leader, though I'm not perfect (no leader is). I'm a man of many failures, but I've gotten many things right — and I credit, to a great degree, my training. In the military, we develop realistic war-like scenarios to ready us for the real thing. We execute training scenarios of different types and magnitudes. My body and my brain remember it all, like panting heavily after a training scenario where soldiers were killed (fictitiously) and where things went horribly wrong. The lessons of those failures were etched in my brain forever as guide rails or like a coach reminding me of those lessons in times of crisis. Throughout decades of leadership, I have pulled lessons from readings and classroom scenarios that helped me (and help me still, to this day) handle any crisis.

When Devoe was killed that horrible day in March 2009, I turned to my experience in Iraq 2006-2007, where we lost numerous soldiers. I turned to training scenarios in a career (at that time) approaching 20 years and I turned to readings I had done on combat leadership. I hope this book can help serve that same purpose for you.

Trust Your Gut, Do Your Homework

At the moment of crisis, when decisions are required, many leaders rely on gut feelings to guide them. This phenomenon is, however, more than just a sensation in one's gut. The gut is not totally ignorant. It is, indeed, informed and guided by one's personal code, experience, and training.

The gut, unfortunately, is not always correct. You probably recall a few times in your life where your gut steered you wrong. In those cases, it was probably guided by ego and emotions instead of reason, principles, values, and experience. Your gut needs this book! You won't think of a specific chapter or page number when crisis hits. You may remember (but will not have memorized) the concepts presented in this book. And over time, the concepts might become fuzzy, but they will guide you, nonetheless ... if you pay attention now. As you build, grow, and prepare your organization for its mission and for the future, you'll rely on the definition of crisis and the "Crisis-Ready Foundation" that I will share with you in the pages ahead. You'll recall how to react when the "Boom!" occurs. You'll emulate the "5 Cs of Crisis Leadership" under duress. You'll form Crisis Response Teams (CRTs) and/or you'll follow the concepts of the TOC (the Tactical Operations Center).

The gut is not totally ignorant. It is, indeed, informed and guided by one's personal code, experience, and training. The gut, unfortunately, is not always correct.

Read this book. Share it with your leaders. Keep it close to reference when crisis visits your team. You'll be a better prepared leader for yourself and your people.

Who Am I to Write This Book?

It's a fair question. These days, everyone has big opinions about leadership and many of us have our own takes on "crisis management," crisis response, or avoiding crisis altogether. So, you might be curious, as you commit yourself to the chapters ahead, who I am and what unique perspectives will inform what I will attempt to teach you. Here's what you should know ...

I bring this book to you with decades of crisis experience from my years in the crisis-plentiful United States Army, and then several years of private, small-business ownership and operational experience in my post-military life. In what I hope to teach you, I will leverage my military experience serving in big, bureaucratic organizations rich with people and resources in a time of war, but I promise to simplify all that I have learned. I won't suggest solutions that require the depth found in a large military organization or a Fortune 100 company, knowing readers will be approaching these lessons and concepts from their own perspectives in a variety of industries and in organizations of all sizes. I'll whittle down large, complex methodologies and systems into something useful for you, the small-to-medium-sized organizational leader or business owner. Because a crisis can happen anywhere and, frankly, a crisis in a small organization can be more devastating than a crisis in a behemoth, where there's more money in the bank, more people to step up to help, and more "staying power" in every single way.

Most businesses these days are lean. Departments, sections, and verticals (and the projects they manage) are typically run by a handful of people or sometimes even a single person. This book is written in the spirit of serving those types of businesses — I wrote it with you and your organization in mind: the entrepreneur and the nimble leader, at the helm of a small business with few people, awaiting its own crisis. What I serve up here is simple, practical, and useable. I've had successes and failures as a solider and small business owner, and I'll deliver those lessons in the pages ahead through real and fictitious stories and examples. I hope you will consume each chapter and embrace and implement its practices right away, if needed; or keep this book handy, like a fire extinguisher to douse the flames of crisis, when the time comes for you and your team (as it surely will). Share this book with your team and you'll be stronger, harder, and ready for crisis, whatever form it takes.

You might be thinking: "But you're a soldier, Rob. What do you know about small businesses?" That's a great question. Here's what I know. After my 27 years in the United States Army, I made the much-anticipated and somewhat harrowing military/civilian transition.[1] I founded Rob Campbell Leadership in 2017 as a solo venture (with a few wonderful assistants and contractors) through which I deliver leadership coaching, consulting, speaking, and training programs. Rob Campbell Leadership is a lifestyle business, meaning it allows me to work from home (except when I'm physically on-site with a client company or an event audience), generating enough of a profit to cover my expenses and put some money in my pocket. I didn't start the venture with an exit in mind. In truth, I had no idea what I was doing. There may be an exit someday, but I believe, at least today, that when I stop consulting, training, and speaking, so does the business. I am the product and, like most consulting practices, that business won't be something I ever sell or franchise; it is an extension of me. Starting, growing, and running Rob Campbell Leadership taught me a ton about business. It taught me about the constant battle for cash flow and it taught me how to craft, deliver and close proposals. It has taught me about marketing and, of course, it has taught me just how impossible start-ups are. I fooled myself into believing people would line up at my door with checks in hand. But in the civilian world of business, nobody was saying, "Sir, yes sir" or saluting me or seeking my advice. In truth, no one knew who I was, let alone understood the value proposition of what I had to deliver. It was a steep uphill climb. I was an amateur entrepreneur at best when I founded my LLC and, because I'm not one to admire a problem, I joined an entrepreneurship program called VETtoCEO to do my part to promote business ownership among veterans and to do a lot of learning of my own.

As years passed and my business acumen grew, I found myself yearning to climb yet another mountain, as I had done in the military. The Army was a goldmine of advancement and opportunity. For me, it started with a 24-man platoon and ended with a brigade of nearly 5,000. I had led a small group and commanded a large enterprise in famed units, such as the 82nd and 101st Airborne Divisions. Leading was personal to me. I relished the opportunity to invest in people — to make a difference and

1 If you or someone you know are planning to retire or depart from the military (or have recently done so), be sure to check out my book about this process. It's a book that was half a lifetime in the making, and I'm very proud of it. Service members and spouses/families are finding it incredibly validating and helpful. Available in paperback and ebook formats. *At Ease: Enjoying the Freedom You Fought For — A Soldier's Story and Perspectives on the Journey to an Encore Life and Career.*

leave behind a legacy. This yearning did not stop upon my transition from the military. In my leadership, I felt this void of having a team around me each day to care for and lead. I wanted something private (not military) to assume control over, scale, and maybe sell someday. I didn't feel like a true entrepreneur managing a significant cashflow, marketing on a grander scale, meeting payroll, and strategizing on how to expand the business into something bigger and better. I wanted an opportunity to buy something that already existed, akin to what I did in the military — to become its owner and leader and make it better. In 2020, I started exploring small businesses for sale, mainly in the trades, which I believed to be recession-proof. I engaged in a few formal conversations and did some due diligence to see if they would be profitable, scalable, and a good fit for me. In 2021, I found one.

Leading was personal to me. I relished the opportunity to invest in people — to make a difference and leave behind a legacy. This yearning did not stop upon my transition from the military.

I connected with a business broker on LinkedIn who asked if I would consider a window-fashion company (a business that sold and installed shutters, blinds, and shades) in Johnson City, Tennessee, where my wife, Leslie, and I had landed post-Army after a few years on the beach in North Carolina. Hmmm ... a window-fashion company. I gave it some thought, looked over a few documents, discussed it with Leslie, and expressed my interest. I was instantly impressed when I visited the business. The building was also for sale, so I would be my own landlord. It had wonderful owners, a very good record of performance and growth, and, by all accounts, a great reputation. The business included a custom picture-framing shop and gallery space, perfect for Leslie, who is an artist. I formed a small team of advisors and put a deal together to buy All About Blinds. I closed on the business in January of 2022 after a long period of time during which I learned the ins and outs of window fashions and raising enough capital for the purchase. I'd bought a business!

Year one (2022) as the owner of All About Blinds was full of ups and downs, of making mistakes and learning from them — of tackling various "mini crises." The aftershocks of the pandemic could still be felt — people still somewhat reluctant to interact and assemble in groups. The time I spent prior to closing on the business showed me that the home-improvement space was still alive and, in many spaces, thriving. My "mini crises" came in the form of misunderstanding cashflow. My budget at All About Blinds was vastly different (and smaller) than a military budget and I confess that I spent (in some cases) as I did in the military — believing there was more than there was. I wouldn't say our very existence was ever in jeopardy nor that I came close to missing payroll. I paid a few bills late but never defaulted on my debt. I burned through some cash, which decreased the resilience of the business to weather storms. While we never ran out of money or had a major crisis, there were some precarious tipping points where I could envision our reputation taking a turn for the worse or our mistakes becoming more costly than we could afford. That first year, our errors occurred in three places:

- Bad measurements
- Errors in placing orders
- Misunderstanding customer desires.

While it is true that mistakes are common in the first year of a business (or at least during the first year of new business *ownership*), I should have seen some of this coming and mitigated it. We learned and grew as a small team of four and overcame our mistakes. Things got easier in year two and year three. But time in business or time at the helm as a leader aren't enough to prevent all disruptions or crises. There will be more errors in my company's future — and even full-scale crises — but, as those "Booms!" arise, I'll practice what I preach in this book to keep the ship afloat and sailing smoothly. I'll lead and manage in crisis. And so can you.

In August of 2023, with my blinds business now humming along and enjoying success, I decided to open an ancillary business in the trades and home-improvement space: a painting business offering interior and exterior all-surface painting. I did this on the back of my blinds business which, at the time, had more than 3,000 customers, many of whom were involved in a home-improvement project of some kind. My intent was to diversify revenue streams. If the blinds business took a hit, the painting business could pick up the slack. I met a fine young man of character with some painting experience and painting business acumen, and we teamed up behind this start-up. To date, we have four teammates and a steady stream of revenue. Painting is not as "cut and dry" as window fashions, so my learning curve is steep. We've experienced some setbacks — crises of different flavors — but I've applied what I share in this book to overcome them. It is early, and we are a tight but newly formed team akin to my military unit in March of 2009. There are no assurances, but the future is promising. My heart and my mind are open to what is to come.

What We Have Lost: A Word on Patrick Devoe and His Death

It's time to turn the page and jump into Chapter 1 of *Left and Right of the Boom*. As you do, I hope you'll remember what you were thinking and feeling when you opened this book to find that my very first story was about the death of my soldier, Private First Class Patrick Devoe. Please know that I didn't share that story as a "hook" or an emotional gimmick, and I didn't introduce you to Devoe for a fleeting interaction. Throughout this book, I'll revisit the events surrounding Devoe's death. I do this in honor of him and to help you learn, to help you prevent failure, to help you lead, and to help you manage in crisis. I share what I know from my own memory. What I share is fairly accurate but not perfect. I don't have the actual investigation and its findings produced after his death, but I recall enough to help you.

I take full responsibility for Patrick Devoe's death. Of course, there were leaders at several echelons of my U.S. Army unit who made decisions leading up to that day, and who trained and prepared their men and women, but I was solely responsible for training and preparing my unit for combat. I approved the missions my men and women undertook. I knew we were in a very vulnerable period — having just arrived in Afghanistan, eager to take the fight to the enemy and make a difference in our year there. There were several things I could have said and done that may have prevented Devoe's death. That's a difficult pill to swallow.

As I look back to that greatest of all crises, I'll share mistakes made and lessons learned but I will not (and do not) pin blame on my subordinate leaders. I carry regret, shame, and sadness. I know I'm not alone. I have learned not to beat myself up. I was not a reckless leader. Collectively, *we* were not reckless leaders. We operated off what we knew at the time. We did our best. We were (and are) human beings in tough, chaotic situations. I'll use "we" throughout the book, as the errors were shared among several leaders, but I take and accept full responsibility.

While his death changed me forever, I feel a great disconnect from Patrick Devoe. I did not know him personally. I know I probably interacted near and with him numerous times, even spoke to him personally once or twice, but I have a hard time recalling those moments. It is the burden of being a commander of more than 400 soldiers. You can't possibly interact and know each of them personally, though I always tried. I regret that — I regret not having forged a personal connection with him somehow. In truth, I feel a sense of guilt using his name at all because I didn't know him as deeply as I desired. I did not know his father, Patrick Senior, or his mother, Susan. I did call his mother from the battlefield via satellite phone to offer my sympathies and do what I could to assuage her grief, but the interaction was brief. I left the real interaction to the leaders below me who knew and loved him.

This book has forced me to return to that period of my life as an Army leader — to relive the horrors surrounding Devoe's death. It is not a place any combat leader wants to return to or a burden we want to carry. While we can't erase the images, sounds, smells, and memories of our darkest moments, we wish we could. I seek no sympathy though. I feel an intense feeling of guilt confessing any hardship I may have experienced over Devoe's death and the return to those days of Spring 2009. I reserve and wish to shift all sympathy toward those who knew and loved him, for it is they who are forever scarred — deeply and personally, far more than I — by his loss.

When I wrote to his mother, Susan, following Patrick's death, I knew there were no words I could offer to ease her grief. I told her instead that, while the Army made him a solider, she gave him his big heart and his caring and nurturing demeanor. It was that love — that selflessness — that made him so special and that made his loss so difficult for us all. We were proud that he served and soldiered among us. We are proud and fortunate to have known him. Each day, in the darkness of the early morning hours, when I make coffee, in a moment of quiet solitude and remembrance, I think of him. It is a simple yet enduring ritual I cherish — a gift he gave me. I offer this book and these lessons in his honor.

During our Afghanistan deployment, we lost another soldier, Nicolas H.J. Gideon, 20 years old from Murrieta, California. Nick, as he was affectionately known (or "Gideon" — customary for his Army chain of command and buddies to call him) was killed in a rocket-propelled grenade (RPG) attack on July 6, 2009. I visited with his platoon mates shortly after his death. They shared funny and flattering stories about him. We bonded in our grief. His buddies — courageous warriors and men of character — actually tried to console *me*. It is a moment I'll never forget and one that I will long cherish. This book is indeed dedicated to Patrick Devoe, and it focuses solely on his death, but Nick's death was no less tragic and no less impactful on my unit and his family. The lessons of love, loss, and leadership — of leading before, during, and after crisis — found in this book reflect Gideon's indelible stamp on me and his unit. For his loss, as with Patrick Devoe's, I take full responsibility. Like Devoe, I think of Gideon each morning.

CHAPTER 1

Crisis

A book about crisis begs an explanation of what crisis is. Therefore, I'll begin this book with clarity, defining and understanding crisis before we travel further. Words matter, especially a word like "crisis" and especially when spoken by a leader. Words like "tragedy," "setback," "crisis," and "scandal" are as important as the crisis itself. If you've read my books, you know me as a definition man; it is from my military upbringing. Definitions strip away the ambiguity around giving directives. They provide common ground and understanding for everyone. They provide the needed clarity subordinates and collaborators seek. Definitions can be remarkably effective in simplifying things for people. And, in times of crisis, clarity and simplification are paramount. Follow the dictionary example when defining things for your people. Here is a definition of the death of Patrick Devoe: "A significantly tragic setback for this organization, which brings momentary pause and threatens the confidence and focus of our unit and its people." At the outset of a crisis, it's imperative that we get down to brass tacks about what happened or is happening and what it means for us all.

As such, I'd like to offer two essential words for definition, which may typify challenging times for your organization — crisis and decisive. When we come up against a crisis, we must — above all — be decisive.

When we come up against a crisis, we must — above all — be decisive.

The Oxford English Dictionary (*OED*) defines "crisis" as *"a state of affairs in which a decisive change for better or worse is imminent; a turning point ... A situation or period characterized by intense difficulty, insecurity, or danger, either in the public sphere or in one's personal life; a sudden emergency situation."*

Heavy stuff — imminent, requiring decisive change.

Merriam-Webster defines crisis as *"an unstable or crucial time or state of affairs in which a decisive change is impending"* and *"a situation that has reached a critical phase."*

Both definitions tell us that a crisis involves impending or imminent "decisive change." In short, when a crisis happens, you can't sit still. A crisis is a moment for decision-making and action.

In a crisis, words and clarity are important for two reasons. First, they shift the thinking of leaders and employees. When words like "crisis" are spoken, they capture (or should capture) the attention of people as if shouting *"Stop!"* or *"Freeze!"* Second, the words chosen and articulated during a crisis could (and often should) initiate some form of significant action or drastic change. When I used the word "crisis" in the organizations I led, my intent was to capture the attention of my people, to cause a collective and momentary pause, and to put everyone on high alert. "Crisis" — the very word — signaled a change demanding everyone's attention. My people knew the stakes were high. We would pull out all the stops. I was — as I recommend you be — judicious with this word, as I knew it could radically alter our course. In your leadership, be serious, intentional, and know the implications and ramifications of announcing a crisis. We must acknowledge and understand that decisive action will likely follow ... which brings us to another important word for crisis situations: *Decisive.*

The *OED* defines decisive as *"that decides (a controversy, contest, etc.) conclusively or finally; determinative."* Other dictionaries use phrases like *"settling an issue"* and *"producing a definite result"* when teasing out the definition of "decisive." I've included "decisive" here in my scene-setting for this book's deep discussion of crisis planning, management, and response because of its relationship to crisis. In a true crisis, organizations and their people will act decisively. They *must* act decisively. When a leader utters the word "decisive," their people ought to know that the work ahead will result in some remedy to a problem, closure to a lingering issue, or something that places the organization on the proper trajectory.

When I used the word "crisis" in the organizations I led, my intent was to capture the attention of my people, to cause a collective and momentary pause, and to put everyone on high alert. "Crisis" – the very word – signaled a change demanding everyone's attention.

In the military, we use the words "decisive" and "decisive point" to signify the action or geographical point upon which success of the mission depends. Consider the example of a military cordon and search operation. You're likely familiar with the term "cordoning off," meaning creating a circle or barrier around something to protect it, like when police "cordon off" a crime scene behind yellow and black plastic tape. A "cordon and search operation" is conducted when a police or military organization suspects someone who must be killed or captured is in a subset of a town or village. Like a fastened belt, the area is surrounded or cordoned before the search begins. It is decisive then that the cordon be complete for the operation to succeed. Many would see the actual search or capture of a bad actor or enemy combatant within the cordon as decisive. However, without a successful cordon preventing escape (which is highly likely in these operations), chance of mission failure increases greatly. The search is important for sure, but the cordon is decisive. When I said "This is decisive" in the Army, my people knew we would give our all to whatever I was labeling as decisive. Nothing was more important. Success hinged upon that which I labeled "decisive."

Success hinged upon that which I labeled "decisive."

How Do *YOU* Define a Crisis?

Using a business example, let's say that a software-as-a-service (SAAS) company is about to launch a new technology service to diversify its offering, making it more crisis proof. The SAAS company should trust that the lines of code were written properly and that the web links to the product will operate so that once the link goes live, all will function as planned. Indeed, the new program requires all the attention to ensure it works; it should be "all-hands-on-deck" to roll out the new software program, no matter the cost. But this product launch is designed to increase revenue so if no customers show up, the program becomes a liability and not an asset. Thus the "cordon" in this case is the shaping effort or marketing campaign preceding the launch. It is decisive then to develop and implement a marketing campaign that gets people "standing in line" to purchase the new service once it becomes available. Function issues with the new product can be overcome. Nobody "standing in line" to buy the new service — no marketplace demand or no way to reach your customers — ensures mission failure.

Both words — crisis and decisive — can lose their intended impact if leaders apply them poorly, thoughtlessly, or too often. Akin to crying wolf, repetitive uses of high-stakes words in a superfluous manner will only weaken them. They become powerless and people won't take leaders seriously. "Uh oh, another crisis! Folks, we have a crisis here." A leader who overreacts when the company is out of coffee, when a vehicle is slightly damaged, or when someone's feelings are hurt (oh, those bruised egos!) diminishes their ability to lead when a real crisis occurs. As leaders and their people navigate any given workday, challenges occur, and bad things happen. A bad customer review might hit the internet, or an employee suddenly quits. Are these examples of a crisis? Perhaps they could be the beginning of something catastrophic, but probably not. The key questions to ask (and answer) are these:

- "How do we define this current turmoil? Does it indicate our eventual demise?"
- "Are we really in a crisis?"
- "Will the business fail?"

Indeed, a single bad customer review followed by several more might put in motion the end of a business. The key is knowing where that tipping point is. Knowing when, as *Webster* defines, the situation has reached a critical phase where decisive change is necessitated.

A leader who overreacts when the company is out of coffee, when a vehicle is slightly damaged, or when someone's feelings are hurt (oh, those bruised egos!) diminishes their ability to lead when a real crisis occurs.

Indeed, there is some intuition or gut feeling that leaders must apply when considering the use of these words. Perhaps the third or fourth bad customer review is the tipping point, especially if those bad reviews allege something illegal or unethical, or if those reviews are from reputable sources or the media. This would represent a trend. In every leadership position I've ever held, this is what I was looking for before making decisions or declaring a crisis — trends. The right type of trend indicates that a crisis is afoot and that decisive action is needed. The resignation of a single employee (especially if they're not your CEO or key leader) may or may not cause a mass exodus of people. Perhaps the second or third resignation among key leaders or contributors signals the tipping point. While both situations — bad reviews or bad retention of key employees — could lead to a company's ultimate ruin, leaders should refrain from crying foul or labeling the problem "a crisis" before it's time. Rather, leaders ought to investigate to determine root causes of these issues and to see if trends exist. Furthermore, leaders must determine if the survival of the organization (or at least the organization as you now know it, with its current workforce, product and service offerings, etc.) is truly at stake. This analysis and choice of words are both fundamentally important. Start here. Don't rush to utter words you cannot take back; don't rush to take decisive actions before it's time.

Battlefield Tip: Before You Call It a "Boom!"

You might be wondering, "How can I avoid rushing to declare an emergency or call a crisis 'a crisis' when I have no time to waste?" That's a great question. Taking time to assess, analyze, and decide is a process that exists on a continuum.

Sometimes, you'll have lots of time. Maybe your organization or one of its programs just lost its accreditation or good standing, but you have weeks or even months before the accreditors will be announcing their decision; in that situation, you have plenty of time to be thoughtful about how you approach the situation, what you call it, and what decisions you'll be making in its wake.

And sometimes, you will have almost no time at all. Maybe your building is on fire — right now. You might have just minutes or seconds to sound the alarms, convey the details and seriousness of the situation, and take decisive action. In that case, it helps to have experience "calling it" during crisis response and making swift decisions. But you still — even when there are only seconds to assess and decide — can and must take a deep breath, draw upon your experiences as a leader who has seen things go "Boom!" and then get moving. There is always, always time. Even when that time is very, very brief.

Was the COVID-19 pandemic — the black swan event of our times — a "crisis?" It certainly was for many millions of people (many who did not survive) and probably thousands of organizations (many that did not survive), but the pandemic was not a crisis for everyone. And it wasn't a crisis right away. Early reporting on the SARS-CoV-2 virus, which causes the disease we came to know as COVID-19, did not necessitate the uttering of "crisis." At first, it was simply "news." Medical news, science news, cultural news, international news. While the news captured the attention of many during the initial weeks, it didn't impact public or private businesses, and it wasn't having much (if any) effect on individuals, families, or communities outside of China, Thailand, Japan, and the Republic of Korea — the countries with the first confirmed cases.

In December of 2019 and January and February of 2020, the situation certainly justified some imaginative problem solving and preparation, but most businesses and hospitals around the globe operated as usual. However, as the pandemic spread and there was no vaccine to prevent serious illness and death among those who contracted the virus, consequences grew more severe. Governments began to act decisively, and others followed. After more than 118,000 cases in 114 countries and 4,291 deaths, the World Health Organization declared COVID-19 a pandemic (a "crisis") on March 11, 2020.[2] Public buildings, restaurants, stores, and event venues were ordered closed. Yet even initial closures, while strenuous, did not result in an immediate crisis for businesses with sufficient cashflow. Several businesses had survived natural disasters and other events, which had halted their operations for a relatively short period of time. Many of us — me included — were thinking, "Wow. All non-essential businesses are closing, and we're all supposed to quarantine at home for 14 days. This is weird. But everything will go back to normal in two weeks. I'm sure everything is going to be fine."

The tipping point came when there was no end in sight — no assurances of when it would no longer be dangerous to be in a crowd and no prospects for when customers would be back. For most businesses, the mandatory and necessary pause of operations was placing an immediate and enduring negative impact on cashflow, the very lifeblood of a business. There was no money coming in, but there was still a lot of money going out — in overhead, employee salaries, etc. Most governments did not designate an end to closures, so business leaders were left in limbo. Some spoke of "flattening the curve" to allow for medical institutions to balance out the number of available hospital beds with the number of sick people needing them, but it was unclear how long that could take. Hospitals, clinics, morgues — they were reaching capacity. And then, of course, the variants of the virus appeared, prolonging restrictions. Even still, some businesses, like restaurants, would adapt

---

2 Centers for Disease Control and Prevention, CDC Museum, COVID-19 Timeline, https://www.cdc.gov/museum/timeline/covid19.html.

— shifting from dining room service to delivery and/or pick-up, and remain in the black (though not by much) on their balance sheet. Some organizations were going to make it — and some truly innovative and lucky ones (like Zoom.us and the companies that manufactured masks and plexiglass partitions) would even thrive. But in the first year of the pandemic, 200,000 extra closures took place in the United States business landscape alone.[3] For many, many people and organizations, this was a crisis of epic proportions.

REFLECTION QUESTION

Deciding and declaring in a global crisis. How did your organization do?

Was the COVID-19 pandemic a crisis for your organization? If so, at what point did it become a crisis and at what point did it stop being a crisis? Take a moment to recall how and when your various crises were announced and what decisive actions were taken. Then think about whether you think your declarations of those crises were made too soon or too late and whether they were handled with decisive clarity or fearful ambiguity. What would you do differently if you could go back in time to lead your organization through it once again?

Crisis or not, decisive or not, tipping point or not, you must make the distinction within your organization. Just because the business next door is in a crisis that requires drastic action doesn't mean the impact is the same on your business. Even global disasters (like a pandemic) have local, varied impact. As such, only you, the leader, can define the situation and decide how to move forward. Only you — backed by your immediate team, confidantes, mentors, and even facts, data, and books such as this one — can determine whether your organization is truly in a crisis

3 Ruth Simon, "Covid-19's Toll on U.S. Business? 200,000 Extra Closures in Pandemic's First Year," *The Wall Street Journal*, April 16, 2021, https://www.wsj.com/articles/covid-19s-toll-on-u-s-business-200-000-extra-closures-in-pandemics-first-year-11618580619.

("Boom!") and if decisive action is required. Make the call. Define the words you are using and the concepts at play — and be clear about your intent for your people by communicating early, often, and thoroughly. If you don't, you'll only add fuel (i.e., confusion) to an already burning fire.

CHAPTER 2

Left and Right of the Boom

The improvised explosive device (often called simply an IED) became the weapon of choice used by our adversaries in Iraq and Afghanistan. The U.S. military found itself unprepared to deal with it. As in business, combat has a way of throwing curve balls, presenting something novel or new — something particularly nefarious, brutal, strange, or unprecedented — that nobody is prepared for, forcing people to adapt, even under fire. In World War II, the hedgerows dominating the bocage landscape (mixed woodlands and pastures) of coastal France impeded the mobility of advancing troops. Hedgerows were walls of thick vegetation and trees, which separated farm fields and livestock. It was like fighting a war in a maze. Air and land forces underestimated the impact of the terrain of the advancing armor and infantry. General Omar Bradley's advance east — following the seizure of the beaches — became particularly difficult and deadly as German forces, using hedgerows to their advantage, fired upon exposed tanks attempting to breech these well-grown thickets. Soldiers adapted by fashioning breeching devices to the front of armored vehicles and adjusting tactics, such as using stand-off weaponry and air support to defeat entrenched forces and maintain momentum.

With new wars came new "hedgerows." The proverbial hedgerows of Iraq and Afghanistan came in the form of IEDs. These improvised bombs impeded movement and pulled attention away from the indigenous population, which the enemy was using as its support base. Using old Soviet-made munitions, like artillery shells left behind by previous wars, or mixing chemicals, such as nitrate found in fertilizer inside compact containers, IED makers would attach detonators to these bombs, hide them, then ignite them in a variety of ways: beneath the surface of roads, in culverts, or even strapped to animals or human beings. As the war progressed, so did the shrewdness of the IED network and cunning ability of its actors. With this acumen came increased casualty numbers. Soldiers were ill-prepared to detect these bombs, react to them appropriately, and survive their blast effects. Like their World War II forefathers, soldiers adapted — lining vehicles with sandbags and scraps of metal, adjusting their tactics (such as avoiding roads and altering routes to detect or avoid these bombs), and expanding their intelligence capability to degrade the IED network.

The Defense Department took heed, forming the Joint Improvised Threat-Defeat Organization (JIDO). JIDO traced its roots to the U.S. Army's Counter-IED Task Force, formed in 2003 as IED casualties mounted. These organizations took a "working *on* the business" vs. "working *in* the business" approach to the IED, studying the network in depth — from the financer to the supplier to the emplacer. This wider, more strategic approach resulted in better tactics, intelligence, and protective equipment. It was from this comprehensive effort that the phrase "left and right of the boom" — which inspired the title of this book — was born. It was catchy and popular, and it spread quickly across the force. The terminology — "left and right of the boom" — was used often by leaders at all levels to get servicemembers thinking about the IED from a wider perspective. Suddenly, we were all thinking about what happens before the explosion ("left" of the boom) and what happens after the explosion ("right" of the boom) ... and, of course, what it might look like to avoid the boom altogether.

Working "on" the business to make the United States military fundamentally more adept at responding to the threat of improvised explosive devices, JIDO brought together experts from across the government and private sector to examine this deadly issue and then to leverage the power, money, and technological advantages of the United States to prevent casualties. Training plans were altered to include IED situational training exercises (which we abbreviated simply as "STX"). Mock, nonlethal explosive devices would detonate on units preparing for war. Training centers replicated the IED network, capturing lessons from units forward in Iraq or Afghanistan for immediate implementation in training programs. New V-hulled[4], mine-resistant vehicles and mine detection equipment emerged. Under this comprehensive effort, coupled with a more effective counter-insurgent approach to warfare, casualties dropped. Reporting from the battlefield indicated that coalition casualties fell substantially from a peak of 126 in the month of May 2007 to just 23 by December.[5] Fatalities from IEDs dropped to an average of 11 per month in 2008. We were saving a lot of lives in the war within the war — the war against IEDs. Reports from the battlefield indicating devices "found and cleared" saw more frequency in the TOCs (Tactical Operations Centers) across Iraq and Afghanistan. Our enemies adapted as best they could to outpace our technological and tactical advantages. Indeed, even the most comprehensive effort would not bring casualties to a halt, but under the auspices of the "left and right of the boom" approach, servicemembers stood a better chance of preventing, avoiding, and/or surviving the effects of IEDs.

Suddenly, we were all thinking about what happens before the explosion ("left" of the boom) and what happens after the explosion ("right" of the boom) ... and, of course, what it might look like to avoid the boom altogether.

4 A "V-hull" is a type of vehicle armor designed to increase the survivability of an explosive blast under a wheeled vehicle, deflecting the force of the blast away from the vehicle instead of up through the middle of it.

5 "Iraq War Troup Surge of 2007," *Wikipedia*, As of Last Update on March 7, 2024.

Thinking strategically about all that happens left and right of the boom can measurably reduce the impact of the boom itself.

Left of the Boom

"We have to get *left* of the boom," uttered JIDO professionals and battlefield commanders, for it was here where — before the IEDs exploded — that lives could be saved. Getting left of the boom meant investing time and training in the relatively calm space before such explosions. It was here that leaders and their organizations were trained (and drilled ad nauseum), hardening them for the impending crisis of the IED. Units would practice tactics in training to avoid IED-prone routes and train on IED reaction drills — practicing the actions they might take following real explosions to save lives avoiding secondary explosions, reacting to attacking enemy, and combat lifesaving. Left of the boom also meant a deeper study by newly formed groups of intelligence professionals. These groups would dissect the IED network, tracing these bombs backward to the supply chain and to their funding streams, attempting to sever the chain long before the placement of a bomb. Units invested more time to counterinsurgency training, attempting to shift the tide of Iraqi or Afghan public support away from the IED network toward their own government and security forces. Training centers underwent a transformation and began to closely replicate the urban and rural landscapes of Iraq and Afghanistan. They hired US citizens of Iraqi and Afghan descent to serve as "role players" in training. We did everything we could to replicate the horrific, deadly threats in safe "practice fields" so we would be ready for what awaited us on the battlefield.

REFLECTION QUESTION

Prevention is expensive. Are you investing in it?

No matter what kind of industry, business, or functional area you work within, there is always a crisis looming. Maybe disaster in your industry involves human lives (this is the case in everything from healthcare to aviation) or maybe it involves financial ruin (such as is the case in volatile industries, like investment banking, or in businesses with fickle customers (like the restaurant industry, where diners staying home during a thunderstorm results in a lot of expensive food going to waste). Maybe your "booms" are related to staffing shortages, technology glitches and outages, or public relations nightmares.

Ask yourself: "Are we doing enough to invest in prevention? To practice our response to crisis? To get *left* of the boom?" If your answer is anything less than a confident "Absolutely, yes," it's time to slow down for a moment on the "busy-ness" of the day-to-day operations of your organization — of working "in" the business — and focus some energy and resources on working "on" the business so you can avoid crisis and mitigate the fallout when it becomes inevitable. What can you do — right now — to commit yourself, your team, your budget, and your organization to getting "left of the boom?"

As in business, the military found that there was rarely an overabundance of cash and people to support the "left of the boom" space. Combat-bound units innovated and adapted, using existing resources and knowledge, often ad hoc, to prepare for crisis. Leaders searched inside their own organizations for people who brought aptitudes beyond their military occupational skill (MOS) to assist in the effort. Cooks and mechanics with college degrees or experience in law enforcement and even farming formed what became known as Company Intelligence Support Teams or COISTs, not typically found at lower echelons, to be the brains advising, steering, and altering tactical activity against this threat. In training, commanders developed and tested concepts designed to help organizations survive the crisis. They communicated with units in Iraq and Afghanistan and even traveled to the battlefield to conduct site reconnaissance, bringing fresh lessons back to their organizations for implementation. While training for IEDs was grueling, no lives would be lost during this practice and preparation work. And the work was vital because it saved lives — more than 1,000 a year. Units could pause whenever appropriate to review actions and learn from mistakes, then repeat scenarios and reaction drills. With or without the budgets to do this work as we'd like to, we pressed on ... because the consequences were human. Not focusing ourselves "left of the boom" meant losing our fellow servicemembers. Doing anything other than our very best to prepare was a deadly choice.

There was rarely an overabundance of cash and people to support the "left of the boom" space.

In 2009, stationed in Afghanistan, my unit — 1st Squadron, 40th Cavalry Regiment, Airborne — drilled heavily in this space. We delved deep into our units to find talented, intelligent, analytical soldiers and formed our own COISTs. Even in this left-of-the-boom space, we paid strict attention to the requirements of right-of-the-boom, securing IED blast sites for examination, conducting after-action reviews (AARs) to capture lessons from enemy and friendly actions, and from leader decisions and tactics — information we needed for formal investigations and documentation following IED blasts. We wanted to win, to achieve the advantage over the IED network long before the burying and blast of a bomb. We wanted to save lives and defeat the network without firing a shot. Of course, no matter how successful we might be, the IED would not simply vanish and there was no way ensure that no deaths would occur. However, without our relentless and significant left-of-the-boom focus, we would assuredly be doomed.

Right of the Boom

While the military grew more astute at interdicting and preventing IEDs and became better prepared to survive their effects, our adversaries adjusted their game. They would devise craftier techniques for emplacement, such as off-setting pressure plates to trick military convoys. They introduced newly developed bombs, like Explosively Formed Penetrators (EFPs), to penetrate armor. And they began aiming explosive projectiles at different angles. IED networks would become more clandestine — avoiding cell-phone communications and diversifying their supply chain from old Soviet munitions to fertilizer and radio parts. The IED would not disappear completely. But, with the help of JIDO, after the detonation of an IED, units would get "right of the boom," applying forensics to learn more about the bomb, its origin, and enemy tactics. Even in the chaos of combat, commanders and their units would pause to study and learn from IED explosions. Mini laboratories were established on the battlefields of Iraq and Afghanistan, manned by former FBI agents and scientists who could boost efforts in the right-of-the-boom space. The military formalized its investigative processes and became exceptionally good at post-blast analysis. Leaders would enforce this deep study and reflection on enemy actions and friendly-unit actions, sharing actual vignettes and attempting to trace the explosion backward to its creation. Conclusions and lessons from investigations were shared across the force and brought to the left-of-the-boom space so units could alter tactics and apply pressure appropriately on the IED network. Lessons fresh from the battlefield would be exported to home-station training centers to help units preparing for war. While the left-of-the-boom space received greater emphasis, it was informed and guided by work in the right-of-the-boom arena. This is because we knew IEDs were imminent.

REFLECTION QUESTION

Step Right. What Are You Learning?

Time to get your wheels turning. As you progress through the book, start thinking now of what you *can* or *could* do in the right-of-the-boom space after your next crisis.

Translate the lessons of the IED to your organization. Define the crisis, then ask, "What lesson(s) did we pull from that boom? How can we capture them and place them in the left-of-the-boom space, so we're better prepared for prevention or response going forward?"

Taking It in Phases

This work — of focusing left and right of the boom — took discipline and, moreover, a phased approach to helping leaders and units focus and prioritize efforts amidst a demanding schedule. Phasing is used in military operations to help commanders visualize activities in space and time and to focus people and resources. Phase I of a parachute operation, for example, would be the preparation of individuals and equipment. It would start at a designated time well prior to the actual jump and end when paratroopers touched down on a landing zone. Phase II would begin once all paratroopers were accounted for and assembled in units for follow-on operations, such as seizing designated enemy-held objectives or expanding an airstrip for additional forces (the purpose of the parachute operation). Phase III — the return of forces to home base or the beginning of a newly designed mission — might begin when additional forces have arrived, or enemy-held objectives are seized. Each phase has a defined beginning and end, priorities for activities, resources such as fuel and funding, and guidance from leaders. Phases can and should be announced to every member of an organization and monitored to ensure goals and effects are met. Phases can and do often overlap.

Left and Right, and Round and Round: Crisis Management as a Circular Concept

"Left and right of the boom" is a phrase that, when written, is flat and linear; the process is anything but. In practice, this concept is circular — a continuum that effective leaders recognize and leverage to steady their teams for rough waters. Activities in the left-of-the-boom space are informed by lessons learned in the right. Actions and processes that worked in the past — prevented the boom or softened it — are formally discovered in the right-of-the-boom space. But those insights are useless if they aren't plugged back into a virtuous cycle of preparation and response, ensuring that what we learn "right" of the boom is soon part of our "left" of the boom preparedness. What worked in the past should become standard systems, practices, and drills for organizations.

I always found it helpful to start linear, but I knew I had to force a circular ecosystem of learning and preparation to permeate the organization. I had to develop branch plans, form small study groups to dig deeper when we came across a wicked problem that seemed unsolvable, and/or seek external consultant help. Sometimes, I have found myself alone as the leader — purposefully removing myself from the pace and demands of the moment — dedicating myself to thinking of this process in terms of a creative mosaic versus a scientific, straight, flat line.

The use of the word *"art"* in the subtitle of this book was not an accident. Science was, of course, the other word choice — the art of leadership or the science of leadership. But, as a leader, I was always drawn to the *art* of leadership and operations over the science. Science is easy to find. There are plenty of books and websites where you can learn the science of something and apply its lessons. Moreover, science, in my assessment, is more defined — drawing upon knowledge and testable explanations to arrive at concrete solutions. Art — creativity — is where I thrived. Prior to 9/11, military training and operations (and, thus, military leadership) were dominated by science: know the rate of march of your enemy and the capabilities of his equipment, then mass the effects and military advantages — intelligence collection, weaponry, infantry, and armor — in a synchronized scientific fashion to destroy him. When our adversaries in Iraq and Afghanistan presented themselves, they came with no science. They wore no standard uniform and had no "rate of march." They used the IED instead of the tank or combat aircraft. The IED was a very nebulous weapon system — a, messy, dangerous, deadly, terrifying form of art. We had to become artists — mixing concepts clearly displayed on a palette yet applied in creative ways to form creative and innovative solutions as we navigated left and right of the boom.

Indeed, effective leadership is a blend of both art and science. The science brings together the theories and applications of influence, behavioral economics, and the psychology of human logic and decision-making, among many others. The art involves the application of leadership specific to individuals — what motivates and demotivates Walter or Casey and what may be happening in their work and life that informs (or at least *should* inform) how leaders lead. I know leaders who, if you presented a problem to them, would steer you to a book or theory — science. Others, like me, will tell you a story — art. Both are required and I honor both in leadership and operations; however, in today's wildly unpredictable environment, I've found my place with the art of leadership. In crisis, it has served me well.

Phasing for a business might look like focusing your efforts in the four financial quarters throughout the year, which are defined by quarterly activities and priorities. An annual enrollment period for employee healthcare insurance would be an example of a phase within a phase (a predictable period within a financial quarter). Prior to this period, a company would conduct a "preparation for enrollment" phase. Here, leaders would strategize and plan for the "enrollment" phase. Administrators would focus on getting efficient at creating and closing policies. Practice sessions with salespeople and external underwriters would occur. Leaders would create and inspect policies and procedures and ensure the right people were in the right positions and were prepared for the enrollment phase. The preparation phase would end with final checks, rehearsals, and even pep talks. Leaders would be ready to bring preparation-phase activities, such as training and policy adjustments, into the enrollment phase if needed but they would attempt to leave the preparation phase behind as they entered the new phase. Phase closures and beginnings would be announced formally by leaders, including social events to boost camaraderie, increase understanding, and motivate teams.

In the military, left and right of the boom were, indeed, phases and leaders needed to see them that way and, moreover, lead their people through them thoughtfully and with oversight and discipline. Left- and right-phase activities would overlap as needed while we dealt with the preparation and the learning from a variety of booms.

REFLECTION QUESTION

It's Just a Phase ...

What might "phases" look like for your organization? Do you currently think about your work in phases? How are the different phases separate and distinct? What would mark the beginning and end of them? And how might you overhaul your approach to phases to better serve your internal and external stakeholders, and even to prevent or mitigate the risk of crisis?

Knowing that phases do, indeed, overlap and that conditions change, think about phasing rather than thinking of your work as a never-ending wave with no time to come up for air or make adjustments. Phased work is a wonderful way to conduct operations — big or small.

CHAPTER 3

Establishing the Crisis-Ready Foundation

For a building to withstand a crisis such as wind, flooding, and even an earthquake, builders focus first on the foundation. When building a home near the coast or bodies of water like rivers, lakes, or oceans where millions of people choose to live, engineers generally have two choices when it comes to foundations: stem-wall or monolithic slab.[6] Monolithic means "all in one pour" so the foundation is constructed in one single pour of concrete, upon which load-bearing walls are placed. Monolithic slabs are faster to construct and cheaper than their stem-wall counterparts. They save money for homebuyers up front but don't stand up well to changing conditions (like shifting soil) or crises (such as flooding from severe storms like hurricanes — an all-too-familiar occurrence in coastal areas). The single slab cracks under pressure, compromising walls, windows, doors, and roofs. The bitterness and effects of poor

6 Sposen Signature Homes, "Differences Between Stem-Wall and Monolithic Foundations," May 2023, https://sposenhomes.com/differences-between-stem-wall-monolithic-foundations/.

quality (monolithic slab) remains long after the sweetness of low price and convenience is forgotten.

So, what's a "stem-wall foundation?" Stem-wall foundations are constructed in a multi-part process where a footer is poured at virgin ground level, then blocks are laid, forming a wall that extends up to the finished slab elevation. Fill dirt strengthens the final foundation upon which the home may be built. Unlike the monolithic slab alternative, stem-wall foundations, while more expensive and time consuming to construct, stand up better to changing soil conditions and even flooding. Walls, windows, doors, and even roofs remain uncompromised when ground conditions rapidly change — as is often the case in coastal areas. Now, think about foundations as a metaphor and you'll notice the similarities between actual foundations and the strategic, financial, structural/organizational, and leadership "foundations" of businesses. Choose or tolerate one kind of foundation and your enterprise is less likely to "hold up" in times of crisis; invest in creating and maintaining a better business foundation and you'll be able to "weather" sudden shifts in market conditions, decreases in demand for your products or services, and considerable changes in your company, like a new initiative or a new leader at the helm.

By the time I sat down to write this book — the third book in the *Battlefield to the Boardroom Series* — I'd had the time to parlay my 27 years of military experience into the crisis-rich world of civilian business. I learned quickly — as I'm sure you have too — that the winds of change often come without warning, and that nearly every type of business is full of vulnerabilities during times of crisis, scandal, or business disruption. Regardless of your industry, there will be forces outside your control that will knock you off your foundation — like a new government regulation, a natural disaster, or even a global pandemic like COVID-19. Those forces can up-end the demand for your products or services, change the needs and expectations of your customers, or get in the way of your ability to produce and deliver what your stakeholders need (when, where, and how they need it, and at a price they can afford). A series of bad customer reviews gone viral on the internet can wreak havoc in an organization, and the new, toxic boss with a penchant for behaviors that make headlines (and not the good kind) can throw an entire company into a dangerous quagmire.

The winds of change often come without warning, and nearly every type of business is full of vulnerabilities during times of crisis, scandal, or business disruption.

Crisis is inescapable in Company ABC on Main Street — just as it was for my Army unit. And while crisis impacts entire organizations, the toll is ultimately human (even when it doesn't involve body bags). When crisis happens, *people* — up, down, and across the organization (and sometimes outside of it, in the realm of the customer) — feel the brunt of the impact. People suffer and pay the price. People get inconvenienced. People get hurt (or worse). Some employees of organizations in crisis may be forced to leave as a business struggles to meet payroll. Some may resist the new initiative or other change presented by company leadership. And the fall-out always, inevitably, comes down to the foundation that was laid years or even decades prior. Employee loyalty, engagement, teamwork, and company climate and culture are the walls, windows, and roofs that collapse under the pressure of crisis if the foundation below isn't rock-solid. When the world goes "Boom!" and no one has been minding the vital work of "left of the boom" — and no one has taken the lessons learned "right of" the *last* boom and applied them to the prevention of this *current* disaster — well, it's going to get ugly.

Building Your Crisis-Ready Organization, from the Foundation Up

Only the strongest organizations (and sometimes the luckiest) survive crises, and this strength is derived from a solid foundation. Here's how to assemble the key ingredients for a crisis-ready organizational foundation; each ingredient matters, and they all work together to create the strength you'll need. Leave out one of the four ingredients, and it's like mixing concrete without cement, air, water, sand, or gravel.

Does your organization have what it takes?

The Crisis-Ready Foundation

1. A noble and inspiring **cause,** captured in a guiding and enduring **vision** statement

2. Espoused **values,** such as teamwork, trust, and innovation

3. A **culture** replicating the strong bonds of a family with **inclusion**, **belonging,** and psychological **safety**

4. Good **leadership.**

It is only when these ingredients come together in harmony that a business stands a chance for survival. The same held true for my organization — for the 1st Squadron, 40th Cavalry Regiment, Airborne — when it came time for us to survive the impact of trauma on that dreadful March 2009 morning when Private First Class Patrick Devoe was killed. Our foundation (thankfully and purposefully) was strong. It was shaken, but it would continue to support the structure above. That foundational strength gave me confidence as we moved forward — in our grief and in our mission.

Foundational Element #1: A Noble Cause and Vision

A cause, as defined by *Oxford Languages,* is a principle, aim, or movement that, because of a deep commitment, one is prepared to defend or advocate.[7] My organization, back in 2009 and for the entirety of my 27 years in the U.S. Army, was serving a noble cause: the defense of the nation. The soldiers who I led during my second combat deployment were volunteers serving their country at the peak of our war in Afghanistan. This cause inspired us to wake each morning and give our all; we were devoted to the *cause* (to defending the United States and to countering the insurgency of the Taliban) and to *each other.* It filled us with enduring pride.

We captured this cause in our vision; we wanted to be "individuals and a team at the peak of readiness in service to our nation and each other." That readiness came in many forms:

---

7 "Cause," *Oxford Languages*, As provided by *the Google Dictionary Box*, Search Made in March 2024.

- Ready to take on a committed enemy
- Ready to handle the rigors of combat
- Families ready to manage at home without us.

Author and career coach Jane Downs states that, "When people are serving a cause, they are acting in ways that are beyond personal concerns and direct personal gain; they are at work on a vision and commitment that lights a flame which draws them into action."

Many businesses have a cause, but few promote it. According to entrepreneur Sir Richard Branson, "A business is an idea to make other people's lives better." The idea is the cause.

Many businesses have a cause, but few promote it.

A technology company for which I consulted had built this vision statement: "Empowering communities and building strong relationships." Their cause is to serve underserved communities with internet technology. Parents who choose to home-school their children can be empowered to do so through the work of this company and its people. People with disabilities and with loved ones to look after can profitably work from home because of the service this company delivers. It was never about the technology, but always about the *cause*. The fire that drove this company's people to action was ignited by promoting that cause and vision with their people and providing concrete examples of it.

A medical company I work with has developed a vision to "Advance healthcare and empower nurses." They perform a specific medical procedure and were experts at it, showing hospital administrators and doctors the power and talent of nurses. Their daily work underscored a cause for nurses, oft unsung, to advance healthcare or provide a greater level of care for patients.

What's Your Cause?

Every organization should have a cause — a principle, aim, or movement that its people are devoted to upholding or achieving. What's your cause? Perhaps your business creates a product that is better for the environment than the alternative or provides a service that eases the life of its customers. Maybe your business brings a community together or supports a charity with a portion of its profits. These are all causes ... and people who believe in them will commit themselves even in the toughest of times. If you were to ask your leaders and your employees to articulate the organization's cause, would they all immediately know what to say? And would they all say the same thing? If not, it would be wise to invest some time identifying your cause; then, it's time to get busy promoting it to everyone.

Closely tied to an organizational cause is its vision. Vision provides purpose and direction. *Oxford Languages* defines this type of "vision" as the ability to think about or plan the future with imagination or wisdom.[8] Vision statements are not one-time goals — like being number one in a market or becoming a $100 million business. They are visuals of the business in a future state doing what it was meant to do at the peak of its ability. For my Army organization, we could all visualize the perfection we were striving for. We could visualize ourselves as confident and ready to serve our cause. We committed to the achievement of this vision each day and pushed away distractions, even enemy activity, which pulled us off course. This allowed us to persevere after the loss of one of our brothers.

In my small business, where we provide window fashions and home painting, we formed and committed to a vision of serving: *Serving our people* (our employees), *our customers*, and *our community*. We all put our belief in it. At my company, All About Blinds - Johnson City, we believe that taking care of each other in a spirit of selflessness and teamwork (two of our values) allows our team to perform at its peak.

---

8 "Vision," *Oxford Languages*, As provided by *the Google Dictionary Box*, Search Made in March 2024.

Vision statements are not one-time goals — like being number one in a market or becoming a $100 million business. They are visuals of the business in a future state doing what it was meant to do at the peak of its ability.

We believed at the outset (and still believe today) that through a commitment to service — a commitment that I find lacking greatly in today's environment — we would excel. We believe that the business could be a great vehicle to support important causes, such as promotion of the trades, supporting veterans, supporting local schools, and supporting other community programs. And those strongly held beliefs are being borne out every day.

The cause and vision of both these organizations — of the military units I've led within the Army and the team I currently lead in my home-improvement business — were our guideposts ... our North Star. My leaders and I talked about them constantly. Our people we were guided by and motivated by them. They would walk through fire for them. Unfortunately, however, many businesses spend little time talking about this foundational ingredient —their vision — and in some organizations, there's nothing to talk about because there is no vision at all. And you're doomed either way — whether an organization's vision is missing altogether or forgotten (you all know the kinds of leaders who spend time forming a vision, but leave it behind, tucked away on a document somewhere, or in a dusty "strategic planning" binder, or several clicks deep on a website never to see the light of day). This approach only weakens the foundation.

I dedicate a good portion of Chapter 4 in my first book, *It's Personal, Not Personnel: Leadership Lessons for the Battlefield and the Boardroom*, talking about vision. Without it, an organization is rudderless, drifting on the open sea, letting the current (or crisis) guide them to where they don't want to be. This willful blindness of the future — or lack of a guiding beacon — only amplifies the effects of crisis when it occurs. Rudderless, leaders turn reactionary, taking their eye off the organization's vision. Their decisions become reckless and even damaging and the effects of this are felt below, among people who look to leadership. The immediate and constant flow of information during a crisis overwhelms leaders as they try to process information in rapid fashion. New and challenging demands cloud over their thinking and impact their ability to lead. In combat, we called this the "the fog of war." Our enemies would attempt to overwhelm us with activity and information, then take advantage of our confusion. But no matter how bad it got, our cause and our vision served as beacons to guide leaders through the fog. The same applies to a business.

The immediate and constant flow of information during a crisis overwhelms leaders as they try to process information in rapid fashion. New and challenging demands cloud over their thinking and impact their ability to lead. In combat, we called this the "the fog of war."

In day-to-day activities but especially during crisis, organizations need someone and something to light the way. Visions do this. They provide that proverbial lighthouse that everyone can look to and move to. And visionary leaders teach us to look for the light. The more an organization's vision is promoted, the greater the chances an organization overcomes crisis. Of course, there are no absolutes. A crisis may, in fact, justify a redesigning of a vision, maybe even surrendering to the crisis altogether as a prudent step. By way of example, consider a software-as-a-service (SAAS) company. If the government passed a law regulating the offering of a specific web-based service, the sensible thing might be to surrender and redirect that passion and energy in a different direction. Conquering government regulation and bureaucracy might just be beyond the human, structural, and monetary capability of an organization. There is, perhaps, no sense dragging people through that mud and bankrupting yourself and others.

Crisis or not, a vision must exist. Where there is none, an organization is a car without a steering wheel.

Reader Resources
A Tool for Crafting Your Vision Statement

Need help crafting or recrafting your vision statement? Enjoy this practical tool, in which I offer a simple yet effective way for you and your team to craft a vision statement.

Foundational Element #2: Espoused Values

The noble cause and vision of my Army unit would not be enough to "hold up the house." We needed to add to it a set of values to which we could all commit and espouse in all that we said and did. Fortunately, we had the Army values: *Loyalty, Duty, Respect, Selfless Service, Honor, Integrity, and Personal Courage*. These values are steeped in tradition and history. Soldiers would learn them in their initial entry to service and uphold them throughout. Those who did not were ousted from the ranks — the Army values were (and still are) that important. It should be no different in a business. Here comes another definition ... Ready?

Dictonary.com defines a company as several individuals assembled or associated together; a group of people or an assemblage of persons for social purposes and companionship; fellowship; association.[9] Guiding companionship, fellowship, and association ought to be a set of values. Values are an important foundational element and I share more about values in my first book, *It's Personal, Not Personnel*. A 2019 *Forbes* article "Rethinking the Value of Core Values," highlighted:

> "In any community (or company), values are extracted, lived, and felt — not scripted. They come from what is shared and often unwritten; they create identity and belonging; and, together, they act as

9 "Company," *Dictionary.com*, Search Made in March 2024.

> a compass. While they may also be aspirational, they don't start from a list of sanitized terms or hipster buzzwords."[10]

Ask anyone who has served in the Army to talk to you about the practical application of the seven values, and they'll tell you powerful, first-hand stories of the values brought to life. Call an Army value a "buzzword" to a soldier and expect a fight.

Why do values matter in every type of organization? John Coleman, an author and contributor to *Harvard Business Review* and *Forbes*, argues, "A company's values are the core of its culture ... and offer a set of guidelines on the behaviors and mindsets needed to achieve that vision."[11]

Consider the values of teamwork, innovation, and integrity when thinking of crisis in a business. When the pressure is on and the lives and livelihoods of people and the business are at stake, the need for people to espouse these values multiplies. In fact, extreme levels of selfless teamwork, innovative problem-solving, and brutal honesty strengthen the foundation upon which a business stands. As I think about which values matter the most — left and right of the boom and when we're deafened and wounded by the boom itself — I keep coming back to the importance of honesty. When crisis struck during my Army career, whether in training or combat, I demanded honesty from my people. They had to deliver unpolished bad news to me. We had to bond and count on each other as never before.

Consider the values of teamwork, innovation, and integrity when thinking of crisis in a business. When the pressure is on and the lives and livelihoods of people and the business are at stake, the need for people to espouse these values multiplies.

10 Steinhorst, Curt, "Rethinking the Value of Core Values," *Forbes*, October 17, 2019, https://www.forbes.com/sites/curtsteinhorst/2019/10/17/rethinking-the-value-of-core-values/?sh=7ed73c98a042.

11 Coleman, John, "Six Components of a Great Corporate Culture," *Harvard Business Review*, May 6, 2013, https://hbr.org/2013/05/six-components-of-culture.

A company with teamwork as one of its core values — embodied in the words and deeds of its people — will withstand a crisis better. Members will step up for their teammates. They will challenge and assist other teammates as they collectively attempt to overcome the challenges supplied by a crisis. Finally, an organization that has innovation as one of its values stands a better chance of overcoming crisis. A crisis often highlights the flaws and inefficiencies in an organization, and these require innovation, often on the fly, to overcome the hurdles before you. Through the lens of crisis, consider what your organizational values *ought* to be. Have you currently got it right, or is it time to establish new values? (Right now, while the going is good and you don't have a crisis gobbling up your attention.) In my organizations, when crisis befell us, there was no time to develop, promote, and instill values. It's too late. The values that would save us needed to be developed, communicated, and championed by all — long before the crisis — if the "house" were to remain intact.

Foundational Element #3: A culture of Inclusion, Belonging, and Psychological Safety

To be effective and crisis-ready, businesses ought to be thought of and treated as families. Think about who you turn to in times of crisis. A strong family will immediately unite in a time of crisis. As is often said, family is all we have. When tragedy occurs and all is lost, families can be counted on for moral support and even pooling of resources to help those in need. People operate each day atop a foundation of familial love and support. A business is — or should be — no different. It starts with inclusion, belonging, and safety. Daniel Coyle, in his book *The Culture Code,*[12] offers these three questions to leaders as they assess their own organizations:

- Do we share a future?
- Do we belong?
- Are we safe?"

---

12 Coyle, Daniel, *The Culture Code: The Secrets of Highly Successful Groups*, Bantam Books, 2018.

These questions lie at the core of a healthy foundation for a business. My Army units were no exception. Through hardship, through fun activities that weren't work-related, and because we spent a lot of time together, we became a family. Facing danger bonded us in ways that are difficult to describe. General Stanley McChrystal came the closest when he described the bond as "stronger than marriage vows."[13] There is no stronger foundation than that.

Crafting a sense of belonging begins with the very first interaction between organization and individual — at the interview of a new employee.

While your business may never assault a beachhead, these bonds are no less important and, under the right leadership, they are achievable. These unbreakable bonds are born from a sense of belonging. Belonging, according to *Oxford Languages,* is an affinity for a place or situation. It is cultivated through cues, according to Daniel Coyle, which indicate that a person is valuable, is part of a special team, and has the capacity to meet and exceed expectations. Crafting a sense of belonging begins with the very first interaction between organization and individual — at the interview of a new employee. The interviewee ought to experience a sense of family and inclusion in how the interviewer describes the company, the tone that they set, and the many people across the company who get to interview the new candidate or potential "family member." The interviewee ought to yearn for the climate experienced in an interview process. Once hired, the onboarding process is where bonding really happens: a team with members all present to meet the new teammate, a special welcome note from leaders, and company merchandize — like pens, shirts, stationery, and business cards. These are all signs of belonging.

13 McChrystal, Stanley, "Listen, . Learn ... Then Lead," TED2011, March 2011, https://www.ted.com/talks/stanley_mcchrystal_listen_learn_then_lead.

REFLECTION QUESTION

How Does It Feel to Belong?

Reflect for a moment on your own experience in organizations where you have worked and served. Did you feel as if you belonged? What person or process made you feel that way? Was it only moments of belonging or did the feeling endure throughout your time there? Was belonging instant — on day one — or did you have to work to feel as though you belonged to the organization?

Ponder and answer these questions as your own left-of-the-boom preparation. I've always strived to be the leader I wanted to serve under. Belonging continues to play a large part in my leadership approach.

Inclusion, which according to *Oxford*, is equal access to opportunities and resources. Inclusion — offered right away in meetings, events, and to resources — furthers the sense of belonging.[14] Lastly, seeking feedback by asking the opinion of the new teammate regarding what they would sustain or improve in the organization is an excellent way to make someone feel like a member of the business family.

While it is imperative that we feel a sense of belonging and inclusion at our place of work, it's also vital that leaders work to create psychological safety. You may experience what I would call performative or political belonging, which may feel good while not being truly authentic. For instance, there may have been balloons, people, and cards welcoming you to the team. You may even be asked for your honest feedback in meetings, but if the boss is a tyrant, you may feel unsafe speaking up — as if your employment is at risk. Belonging — when it is genuine and deep — fosters psychological safety. Not to be confused with physical safety (which is also

14 "Inclusion," *Oxford Languages*, As provided by *the Google Dictionary Box*, Search Made in March 2024.

important), psychological safety means that people feel safe to give critical feedback and to be themselves at work. The organization and its leaders create and promote a threat-free environment. Members don't walk around each day in fear of being fired; leaders protect them and champion them. This doesn't mean that employees cannot be challenged or that they need not worry about their performance — psychological safety is not immunity from responsibility. What it *does* mean is that the environment helps them feel at ease and motivated to speak up for the betterment of the organization.

When crisis comes, leaders will depend on the honest feedback of people who feel safe in their roles and safe in their teams. Their sense of belonging encourages them to selflessly contribute to problem-solving or to sacrifice for the good of others and of the company. Organizations lacking this safety will reach the tipping point of disaster more quickly, as people will go into fight-or-flight mode, protecting themselves instead of the business, its leaders, and their peers.

When crisis comes, leaders will depend on the honest feedback of people who feel safe in their roles and safe in their teams.

Creating and fostering psychological safety is an enduring duty for leaders. It won't happen organically or overnight. It happens when the boss doesn't lose his or her mind when a subordinate is a few minutes late when reporting for duty or arriving at a meeting. It happens when an honest mistake is forgiven, and the leader consoles a subordinate who is feeling guilty about it. It happens when the leader goes silent in a meeting, awaiting critical feedback from team members, then applauds them for being open, honest, and critical.

Companies that promote a shared future for their people are more crisis-proof than those that do not. If individuals see themselves in the future of the company (or, more importantly, *as* the future of the company), they will go to great lengths to help in times of crisis when the company's future is at risk. Shared futures come from "our" and "we" language. I work hard to use this language when speaking to my teammates. "*We* may take a financial hit if *this* happens." "Ok, *our* cashflow is looking healthy at the moment. My thanks to all of you for achieving that. Along with celebrating that, I want to caution all of *us* to watch for the pitfalls that put a dent in our profits last quarter." Employees must know that if the business wins, they win. There is no future without them in it.

Foundational Element #4: Good Leadership

The final ingredient in the sturdy foundation of a crisis-ready business is good leadership. Crisis demands it. Without good leadership, a business is doomed when emergencies occur. As John Maxwell — author, speaker, and leadership expert — states in his writings and his leadership podcast, "Leadership is influence — nothing more, nothing less." In crisis, people will need to be influenced in several ways — to take on hardship yet remain loyal to the mission to give more of themselves to solve a problem. I could fill the remainder of this book on good leadership — such a huge topic that I hope I'm demonstrating in various ways — but for our purposes here, I'll focus on those leadership characteristics that are essential for crisis management and high-stakes leadership. *Merriam Webster* defines a characteristic as a distinguishing trait, quality, or property of a person.[15] What makes you and your organization *different* from others? Why are you, your people, and your company *better* than others? What makes you *special* or unique? Key, personal characteristics lie at the core of the answer to those questions. For example, maybe your company is special and better because it hires people of character who are (these are the "characteristics") constantly curious, thirsty for excellence, and dedicated team players. Let's focus on five characteristics for our crisis journey. I'll call them the "5 C's of Crisis Leadership." I've seen them at play in my own career — military and civilian.

---

15 "Characteristic," *Merriam-Webster Dictionary*, https://www.merriam-webster.com/dictionary/characteristic.

The 5 C's of Crisis Leadership

Effective crisis leaders will consistently exhibit (even in the most exigent of circumstances):

- Calm
- Communication
- Courage
- Confidence
- Creativity

As you think about becoming the best crisis leader you can be (because crisis is coming for you at some point, no matter how strong your organization is or how lucky you've been), it's wise to spend some time exploring how you can be more calm, communicative, courageous, confident, and creative. Let's break it down.

Calm

This characteristic is not at the top by accident. "Calm" should be added to the job description of a leader leading any organization under duress. Leaders set the tone in their organizations through word and deed — all they say and do. The team takes on and/or reacts to this tone. A leader emitting anxiety through their words and actions will have a negative effect on a team already under tremendous pressure.

Leaders set the tone in their organizations through word and deed — all they say and do. The team takes on and/or reacts to this tone.

I lived on the North Carolina coast for four years following my retirement from the Army. During this period, we experienced three major hurricanes. With each storm, I cringed when I listened to the national news describing the storm's severity and impacts. These news broadcasts were (and continue to be today) perfect examples of how not to lead. I don't want to diminish the seriousness of these storms, or the loss of life and property associated with them. I understand the duty of a reporter to communicate potential threats to their audience but the "left and right of a boom" such as a hurricane is where calm and order are needed most. Sadly, we rarely see that kind of calm on display during the "breaking news" weather reports. If someone describes a hurricane as an "epic storm of biblical proportions" or the aftermath as "the wake of total destruction" standing next to the one of the three trees that was blown over in Small Town USA, they ought to expect their audience to overreact. Wilmington, North Carolina, my former hometown, has had the hurricane book thrown at it, yet it remains intact. The destruction has never been "total" and no one has ever built an ark. Again, I don't want to marginalize the losses of many, but on a grand scale, Wilmington survived Armageddon just fine. The entire coast was not wiped out nor did we experience utter destruction as the news touted. Sensational headlines garner attention but they don't elicit calm behaviors. Leaders in any role must report responsibly for sure, but they must also offer some calm to their people. They are the only ones who can and, therefore, they must.

Much like optimism is a choice, so too is calmness. Leaders are human, after all, and are subject to the same emotions and fears as anyone else. This said, the best leaders — in times of crisis — turn not to extreme emotions but rather to reason, values, priorities, training, and experience (whether the experiences they are drawing upon are their own or someone else's). A crisis-ready foundation and "left-of-the-boom" rigor will equip a leader with the rational ability to perform calmly and effectively in crisis. Crisis does, indeed, demand quick thinking and decisions — and leaders, somewhat subconsciously, will draw upon their training and experience when acting. They will be guided, even if somewhat subconsciously, by their values and personal code of conduct.

Much like optimism is a choice, so too is calmness.

By now, you've become accustomed to me using military lingo and battlefield examples to illustrate universal concepts in the broader, civilian world. So, you'll forgive me for this next bit of terminology. A leader must take steps to display some "Battlefield Patience" when the going gets tough. We used this term in the Army to describe the characteristic of the leader in the group who seemed unaffected by the crisis — who was able to think and lead calmly, allowing others around them some latitude to express their emotions while they remained resolute and calm. Battlefield patience meant not acting injudiciously. It referred to the leader's ability, in the heat of a crisis, to apply tolerance and restraint — to act intelligently, informed, and deliberate even in the span of seconds. Indeed, it took practice to master this crisis-leader trait of battlefield patience, and it could only work if the leader invested time building a competent and confident team. Battlefield patience or calm can be formed in the left-of-the-boom space. I'll speak more to this in Chapter 4, but the point is that calm leadership can be learned. I am no different from any other person. I have emotions and emotional triggers, and I care deeply when things go wrong — especially when they go so wrong that people get injured or killed. Crisis took its toll on me as a young leader and my emotions spiked. It's hard not to have our emotions right at the surface — to wear our anger, frustration, panic, fear, and sorrow on our sleeves. It was only through repetition, exposure to crisis, and by watching other leaders display battlefield patience that I learned and chose to be calm.

Battlefield patience meant not acting injudiciously. It referred to the leader's ability, in the heat of a crisis, to apply tolerance and restraint — to act intelligently, informed, and deliberate even in the span of seconds.

Along with learning from others and experience gained with time, a level of intrinsic leader consciousness is required; leaders need to be awake and aware. Leaders who truly understand their role can embrace calm better than those who fail to elevate themselves to the position they hold. Remember that the position of leader — whether it's a mid-level manager or CEO, a staff sergeant or a colonel — is designed not for the individual who fills it but rather for the organization and its people ... and whether or not those job descriptions say it overtly, calm is a prerequisite characteristic for every leadership role. The CEO is a position created to place leadership — influencing, steering, teaching, coaching, caring, and of course, managing crisis, to name a few — at the top of an organization so that it may survive and thrive. Those given the privilege of the title "leader" will succeed only if they consciously appreciate and understand the awesome responsibility and requirements of that position. I feel this every time I am with my people. As if I'm wearing a new suit that is lighted for all to see, I'm aware of my role and its requirements — with the imperative of being calm being paramount. My consciousness is only amplified when trouble arrives; therefore, I'm able to apply battlefield patience — the calm my organization and its people deserve.

You might be wondering if I practice what I preach (because everything is easier said than done), so let's talk about it ...

I didn't fly off the handle when Devoe died. I didn't berate people, demanding more information or pinning fault. I calmly took in reports and offered my support. My emotions did spike in the moments after his death. Hours of travel by treacherous road separated me from my own headquarters and that parental feeling of wanting to be by the side of my suffering children overcame me. In my moment of personal strife, I asked for helicopter transport back to my unit and I went as far as getting the Aviation Task Force commander on the tactical line to express my feelings and request. He was of equal rank to me and was very sympathetic, offering his condolences for my unit's loss. But he did have an overloaded flight schedule that day supporting a complex battlefield. Our conversation was calm and professional. We were both schooled in battlefield patience.

> "Rob, I can cut a helicopter to you, but I'll have to strip it away from another tactical mission. How can I help?"
>
> "No, don't do that," I told him.

My battlefield patience — my calm — allowed me to apply logic to the situation. I had strong leaders on the ground, who I and others had trained well. I'd seen them perform in training and I knew they would rise to the occasion. There was little I could do for my unit at that moment. We were trained for crises like this and, while I needed to remain abreast of the situation, directing subordinate leaders, I could do that from my current, distant location. Furthermore, the enemy was still active and other units would need the support of lift-and-attack aviation to support them. Taking these assets from other combat units for my own use might cause more casualties. Still on the tactical line with the aviation commander, I confessed that my emotions were flowing. He understood and was gracious. I stayed put, communicated with my leaders via tactical radio and did the best I could under the situation.

I've grown to choose calmness when crisis presents itself. But I never said it was easy.

REFLECTION QUESTION

The Calm Within the Storm — How Do You *Perform?*

How confident do you feel that your team could handle the initial moments after a crisis if you couldn't be there to lead them through it? Think about individuals *and* about the team collectively when you answer this question. Knowing that there are rarely enough proverbial helicopters to get us to "ground zero" of an organizational crisis or disruption in the immediate aftermath — at the beginning of the "right-of-the-boom" moment — how might you better prepare yourself and others for the team to fend for themselves when things blow up?

Communication

Along with calmness, good communication is essential when everything else goes sour. Good, timely, thorough, appropriate, complete communication will save the day in a crisis. Give your people something. Even a message that lacks a lot of substance — that doesn't have all the details and surely doesn't have all the solutions — is far better than silence. Silence would have worse effects. Because I'm such a stickler on definitions, communication is more of a verb or an act than a characteristic. Definition faux pas aside, being a "good communicator" is a required characteristic of a crisis leader. Simply hearing the voice of or receiving a note from a leader can go a long way to bring calm to individuals and an entire organization.

Communication sits at the top of the list of leadership characteristics — like its sibling, "calm" — because of its importance. When crisis happens, people look up to leadership. They ought to get something back, such as some form of communication. Many leaders overlook this or focus inward during emergencies. They may not appreciate that they do have something to offer their people, even as small and insignificant as they may feel it is. The challenge in times of crisis, especially the ones that arrive without warning, is that leaders struggle to find words to say. As surprised as their teammates, leaders may fall silent in the shock of the moment. This was familiar territory in the Army. Curve balls would arrive often in training and in combat, stopping leaders and units in their tracks. Dumbfounded, leaders struggle to find words to inform, direct, and/or calm their people. When these curveballs arrived in the Army, we would use the following methodology: "What happened? What is happening? And who else needs to know?" There is always an answer to each of these questions. Along with this central question, "the SALUTE report" — with details about Size, Activity, Location, Unit, Time, and Equipment involved in an incident — helped military leaders communicate in emergencies.

At the risk of militarizing the modern office, these methodologies absolutely work. They are designed to help leaders communicate in the "fog of war," when reports are rushing in, conditions are rapidly changing, and people are looking for information and guidance. A leader in a company can use these methodologies to help them overcome. By simply stating what just happened or is happening, how they understand it, and who else may need to know, a leader can give something (something vital and even reassuring!) to those who are looking up to them.

At the risk of militarizing the modern office, these methodologies absolutely work. They are designed to help leaders communicate in the "fog of war," when reports are rushing in, conditions are rapidly changing, and people are looking for information and guidance.

The COVID pandemic provides the perfect vignette for our purposes here. Imagine this scenario ... A leader in a restaurant, learning of an approaching government shutdown or dining-room capacity restriction, could communicate the specific language of the restriction or their interpretation of it and how, in their initial hasty analysis, they think it might impact the business. The leader could confess that they are unsure — right now — how to react to this news but reassure their people that problem-solving will commence and further communication will come. That message could be broadcast through several means: an email from the leader, a business messaging channel (e.g., Slack, Teams), or an announcement at a company meeting. Here is an example of what that initial message might sound like:

> "Team, tomorrow at 9:00 a.m., the town committee will vote on whether to restrict dining-room capacity in eating establishments in our city. We may be facing a 50% cut to how many people are allowed in our restaurant at one time. I want all our employees to know about this development, so please help me spread the word. I do not know how we will communicate this to our customers yet and I'll be seeking your input so please tell everyone that, if asked about this, they should share that the restaurant leadership is working on this problem and will communicate something to our customers soon."

The scenario above is an example of what it looks (and sounds) like to use the first recommended crisis-communications methodology, by first focusing on addressing: "What happened? What is happening? And who else needs to know?" The second methodology that I think works as well in your company as it does in the military is what's called the "SALUTE report." SALUTE is an acronym for Size, Activity, Location, Unit, Time, and Equipment. The SALUTE report was designed for military leaders in the thick of a firefight at 3:00 a.m., stressed and dead tired. The report, though made for the battlefield, can easily be modified to fit a business. Formally using a SALUTE format, a business can document actions throughout a crisis for use in the right-of-the-boom space. SALUTE for a business could look something like this:

- **Size** would represent the scale of the crisis. Does it affect all or part of the organization? What is the size of the cost or potential cost in dollars of the crisis?
- **Activity** would represent the description of the crisis, for example, a crisis of market share or crisis of cashflow from the immediate impacts of a particular incident or situation. *Activity* would also describe what is happening or has happened, for example, 800 customers have canceled their accounts, the server has quit and may not come back up for another 6 hours, etc.
- **Location** can help if the crisis is specific to a physical or virtual location relevant to the team. For example, if a certain vertical or department was shut down due to a computer virus, a leader could identify where it occurred and what other locations may be affected.
- **Unit** may represent persons, verticals, departments, teams, or groups affected by the crisis.
- **Time** will help for historical purposes. If an organization can track timings on actions or events of a crisis, it can learn from them in the right-of-the-boom space. As it navigates the crisis, an organization may want to revisit timings to learn and adapt on the spot.
- **Equipment** represents key business equipment, systems, or facilities impacted by the crisis. The impact of a flood on a building is a good fit for the *Equipment* line of a business SALUTE report. For example, perhaps the storage room is

inaccessible, there is no lighting on the stairwells, and/or the phones and elevators are out.

SALUTE is a "spot report" given at a moment in time. It is designed to provide information in a holistic, yet brief, manner to inform the greater team and help leaders make decisions and direct their organizations. Now that we've explored some examples of what would go in such a report, let's peel it back a layer to define the sections of the report more simply.

Size. Monetary cost, number of people (employees or customers), and/or equipment or systems affected. Size is about measurable scope.

Activity. A brief narrative of what is happening. Activity can describe both what is happening and what actions are being taken specific to the crisis. This is a description of "what happened."

Location. Where the event is taking place or where it originated.

Unit. The people, departments, sections, verticals, or teams the crisis is directly impacting or who may have caused the crisis.

Time. Time the report is submitted. Time can also include how much time has elapsed and/or the estimated time when the situation may be under control or resolved.

Equipment. Physical equipment (trucks, computers, printing presses, etc.) — yours or others — affected by the crisis.

SALUTE, because it's an acronym, is easy to remember and it helps a leader or an individual think through its contents and communicate effectively in the toughest environments. It could be reformatted and customized to your industry or business, but it must remain simple so that it can be remembered and used when crisis hits. If an organization wanted to develop a new SALUTE format, they could find an easy-to-remember name/acronym using elements of the business, such as finance, marketing, sales, people, and engineering.

Another way to use SALUTE is to add another letter. SALUTE-P could be used to add *People* as a reporting category (though a thoughtful use of SALUTE in its original form can incorporate all the people considerations in the "Unit" section). A *People* category will force a person reporting on the situation/crisis to consider impacts on people, such as hardship, retention considerations, and motivation. For example, using the example of a computer server crashing, Marsha might send a SALUTE to her boss, adding the vital details that a new group of employees is starting today and will hear about this (and not be able to access their onboarding documents, making for a less-than-productive Day 1 for these new hires). She also might share that the sales team cannot operate without the server. Both facts will help a leader understand, in the fog of a crisis, that people need tending to.

In a crisis — whether you use the SALUTE, SALUTE-P, or other communications framework — it's imperative that you get information flowing. Address your most vital stakeholders as soon as possible, and give them *something*. Even if it's only two lines of a SALUTE report (e.g., what happened and where), it's better than nothing at all. I suggest sticking with SALUTE or some variation of it. It's a time-tested approach used on many a battlefield — and it works. Keep in mind that a higher headquarters — a C-Suite, for example — can (and should) tease this type of comprehensive reporting out of a person or team. If a leader or group of leaders finds something missing in a report, ask for it. Strongly consider SALUTE-P. It can help you and your organization communicate better in a crisis because communication in crisis is paramount.

In a crisis — whether you use the SALUTE, SALUTE-P, or other communications framework — it's imperative that you get information flowing. Address your most vital stakeholders as soon as possible, and give them *something*.

If They Won't SALUTE You, You Can SALUTE Them

What should a leader do when one of their direct reports comes to them with news of a crisis but doesn't have a thorough report ready to share? Maybe you're a VP and a debacle in a department managed by a director has just come to your attention. During your initial "What the heck happened?" conversation with that director, you can use the SALUTE or SALUTE-P framework to inform the questions you ask. Set the stage before you run through these important questions. Be aware that emotions are probably flowing. Using their first name, say the following: "To help you and the greater team, let me ask some pointed questions. I do this only to paint a full picture before I act."

1. **First of all, how big of an issue are we dealing with?** Quickly give me a sense of the **SIZE** of the crisis so I know whether to pull others into this conversation before we go deeper.
2. **What happened?** And what, if anything, is already being done to address it? Walk me through the **ACTIVITIES** that created the crisis and any actions or activities happening in the immediate aftermath.
3. **Where is this all happening?** Help me understand the precise **LOCATION** of the crisis — where it originated and where it has spread.
4. **Which groups are involved?** Tell me which **UNITS**, departments, facilities, teams, sections, or groups are involved or impacted.
5. **What's the timing on this crisis?** When did it happen, how long has it been going on, and how much **TIME** do you expect it to take for us to resolve this?
6. **What equipment has been impacted?** What physical assets have been damaged, disrupted or lost? What **EQUIPMENT** is offline, broken, unavailable, or needing attention because of this crisis?
7. **Which people are involved?** Talk to me about all the stakeholders who were involved in the crisis itself and all

the **PEOPLE** — internally and externally — who will be impacted by the crisis. Who, specifically, needs to be notified about this, and when?

Don't expect to remember these questions when crisis strikes. Photocopy this list or make your own "crisis questions" template.

Communication in crisis is not "one and done." Crises always require overcommunicating. Unfortunately, weak or inexperienced leaders often fail to see this. It is rare that one message from a leader will fully address the fear and angst of their people. Besides, few real crises have short-term impacts. Their effects are more likely to stick around, so communication needs to persist as well. It's akin to a broken bone, which will take time to heal and may never again be the same.

In the aftermath of a crisis (i.e., "right of the boom"), there are many types of communication to pursue. First, there is initial communication — as described above, in our discussion of the SALUTE report. Then there is follow-up communication. People will appreciate initial reports from a leader, but their angst will not quickly fade. The best rubric to apply involves a deliberate commitment to follow initial communication with *more* communication when additional information is gathered or progress is made toward a solution, even if solutions or key steps are not complete. Again, be asking yourself and then sharing with others: "What do *you* know and who *else* needs to know?" This question must nag at leaders (and subordinates), forcing them to constantly communicate with their people. Another rubric involves a plan for communicating after a period has elapsed. The early stages of a crisis call for frequent communication. As the situation matures, however, communication can spread out. In many situations, letting only four to six hours pass between reports is a good rule. As an organization navigates its way out of a crisis, leader communication can spread out. At a minimum, leaders ought to communicate every 24 hours until the organization is "out of the woods," so to speak. Even if no new information has arrived or no progress has been made, a leader ought to say so. "No change, still monitoring" can be something to keep that important connective tissue between the leaders and those being led in crisis. It is rare, though, that nothing has happened if a true crisis is unfolding. A crisis action team may have been formed or certain expenses cut, or the CEO may have commenced a study to examine the crisis and its true impacts. "Right of the boom," there is always a flurry of activity and thinking. Leaders must take time to observe and assess what forward movement has occurred and communicate that to their people.

Not only are the messages and their frequency important while communicating in crisis, who to communicate with is also critical. Leaders must consider other stakeholders — like investors, customers, partners, and creditors, just to name a few. Families must remain at the top of the list when thinking of other-party communication. All organizations have this in common; for every person in the organization, there are several loving and supporting family members behind them. Crisis doesn't just occur on battlefields — it happens on Main Street and, depending on the severity of the crisis, families will come together and help an organization endure. Communicating with families must reside right next to communicating with employees. It is imperative, therefore, that a good family network be established and nurtured. It is no surprise that family communication finds its place in this chapter about "establishing the foundation." Always keep families in mind. When you do, your people will respond.

For every person in the organization, there are several loving and supporting family members behind them. Crisis doesn't just occur on battlefields — it happens on Main Street and, depending on the severity of the crisis, families will come together and help an organization endure. Communicating with families must reside right next to communicating with employees.

REFLECTION QUESTION

Do You Know How to Reach the Families of Your Employees?

Now — before your next crisis — is a great time to be thinking about how to improve your "family network" and your communications planning for instances when you should be reaching out to the families of your employees (or the families of other key stakeholders, like patients or clients). Ask yourself:

- Do we have family contacts (beyond the typical "emergency contacts") for our employees? What information do we maintain?
- How often should we be communicating —in good times — with the family network so they will be paying attention to emails, texts, or phone calls from us during bad times?
- Do we have systems, technology, and protocols in place for sending mass messages to this important stakeholder group?
- Do our employees know how to provide us with updates for spouse, parent, child, or next-of-kin contact information, and are we regularly (e.g., once a year), asking employees to confirm that we have current information on their families?
- Have our leaders taken time to make strategic decisions about what kinds of internal information will be shared with families in times of crisis?
- Who can I "pin the rose" on for this most important effort? HR or some other leader on the team, with whom I can check to ensure these questions are answered and our family focus endures?

> Different types of organizations have different needs when it comes to their "telephone tree" of contacts. Maybe you work at a hospital, where it's vital that you have contact information for the families of your patients in the case of a hospital power outage or active-shooter lock-down. Or maybe you work at a factory or warehouse, where it's imperative that you know how to contact the families of your third-party delivery drivers, vendors, or others who are often on your premises. If you work at a school and have minor children under your care, it's probably important to have multiple levels of family contacts — if you can't reach a child's parents or guardians, you'll want to know how to contact selected grandparents, aunts, uncles, babysitters/nannies, or trusted neighbors. Be thinking about whether it makes sense for your independent contractor agreements to include space for 1099 contractors to share, at their discretion, contact information for emergency/crisis contacts.

When Devoe was killed during that fateful, horrible day in Afghanistan, we already had a formal process for communicating. That said, I still needed to consider what to communicate and who to communicate it with. Formal and informal updates would be produced by watch officers and sent electronically between command posts, keeping leaders and soldiers at every echelon informed. I didn't change our mission, but I did place an emphasis on securing soldiers and equipment, preventing further loss of life, and analyzing the intelligence we gathered. While I could access the watch-officer reports, I always went a step beyond the reports and contacted my commanders to speak with them directly about what actions they were taking. In these calls, I could learn of things not included in standard reports. At every step, I communicated my actions to my boss. I could steer my unit through hardship, give my boss confidence in me and my unit, and seek mentorship as needed. I had learned long ago that, while my boss didn't need blow-by-blow updates, he (like me) operated off information — and it was my job to give him plenty of it.

I shared this same intent with my leaders. "I'll never be angry at you if you need to contact me with an update. I *will* be angry with you if you assume you are bothering me or fail to communicate altogether." I communicated early and often, and I expected my soldiers — at every level — to do the same. None of it was easy, and all of it mattered.

I communicated early and often, and I expected my soldiers – at every level – to do the same.

Whether the work you do is extremely "high stakes" or whether you rarely experience a true crisis in your organization or industry, you will be better armed for the future by adopting the practices and mindsets of calm, communicative leadership ... now. Because when the time comes, your people will be counting on you to have what it takes, and competence during crisis first requires calm and communication. Whatever message you choose, whoever you decide to communicate it to, and however frequently you decide to send messages, just never lose sight of the fact that you will surely fail if you fail to communicate.

Communicate, communicate, communicate!

Courage

Merriam-Webster defines "courage" as the mental or moral strength to venture, persevere, and withstand danger, fear, or difficulty.[16] *Withstand* is the operative word here, because crisis will serve up plenty of danger, fear, and/or difficulty to be endured. It's the perfect environment for the courageous. With or without crisis, courage is an essential characteristic of a leader due to the dangers they face. Danger comes in many forms, and it's required for leadership for many reasons:

- Uncomfortable conversations require courage.

16 "Courage," *Merriam-Webster Dictionary*, https://www.merriam-webster.com/dictionary/courage.

- It takes courage to speak to a group of people (as leaders often do) and/or to correct and admonish poor performance.
- It takes courage to bring change to existing processes where resistance may be met.
- It takes courage to exercise battlefield patience when the boom occurs.
- It takes courage to deliver bad news.

Those who avoid danger, fear, or difficulty — or the environments where these are likely — simply cannot lead.

Those who avoid danger, fear, or difficulty — or the environments where these are likely — simply cannot lead.

While it is quite customary to associate courage with crisis, I want to take a wider view here. We often think of courageous people as the biggest, strongest, and bravest among us. Military special operators come to mind or professional fighters or stunt professionals. Those people do, indeed, have courage but, like all leaders, they have fears as well. Actor John Wayne said, "Courage is being scared to death, but saddling up anyway." Accepting that we all face danger and that we all have fears, let me share this perspective. James R. Detert, the author of *Choosing Courage*, offers an alternative view of courage. He challenges the traditional perceptions of courageous people as being the strongest and most aggressive people, existing in an environment where survival — to avoid being hunted and killed by other animals — is essential. Detert offers, instead, three activities that define courage in contemporary organizations. First, courageous people display openness and humility. Openness and humility, when exhibited by leaders, attract people to follow the leader and motivate them to find ways to help. Second, courageous leaders put principles first. This is best described as "leadership is not a popularity contest," according to Detert. Leaders must stand their ground for what they know to be proper and just. Third, courageous leaders focus on making environments safer for others. This refers to psychological safety — "encourage courage," Detert states.[17] Leaders who create safe environments promote ideas, courage, and creativity from their people.

Indeed, courage is needed to be a vulnerable and transparent leader. While "never let them see you sweat" remains a good philosophy (i.e., display steadfast calm and confidence), embracing vulnerability and transparency is also a wise way forward. There is a gauge, however. Leaders can be too much of one or the other — too vulnerable or too transparent. For instance, a leader who displays too much vulnerability may sow doubt among their followers and the greater organization and flatten any hope people may have. Here is an example: "Oh my, this is bad. I'm not sure how we will ever overcome this." Few would want to follow that leader or have any confidence in them because they were vulnerable to the point of hopelessness. Appropriate vulnerability sounds more like this: "Ok, this is messy. Anyone have any initial ideas for how to recover from this?" Spoken calmly, this leader displays appropriate vulnerability — confessing they don't have all the answers. This leader also acknowledges what people are likely thinking — "this is bad." But this leader also speaks about recovery and involves the voices of everyone who might be able to contribute to a solution.

While "never let them see you sweat" remains a good philosophy (i.e., display steadfast calm and confidence), embracing vulnerability and transparency is also a wise way forward.

Transparency refers to personal transparency. When I think of transparency as an identifying characteristic of a leader, I'm reminded of what General Stanley McChrystal shared in his book, *Team of Teams*: "Share

17 Detert, James R., "What Courageous Leaders Do Differently," January 7, 2022, https://hbr.org/2022/01/what-courageous-leaders-do-differently.

information until you're afraid it's illegal." In other words, accept the discomfort of being transparent. Know and accept you'll be uncomfortable but have confidence (and courage) that it is for the greater good. The truth is that discomfort requires courage. Returning to our gauge, leaders don't want to download everything that's on their mind and dump it on their people. Doing so may only confuse or misdirect a person or a team. For instance, a leader may not want to share their doubts in the ability of a person, team, or organization's ability to overcome a challenge. That might only crush any hopes of perseverance and progress. Rather, a leader can and should fill any knowledge gaps or share any concerns they have, while keeping to themselves any thoughts that may stifle hope. This is leadership vulnerability.

Vulnerability has always represented strength — not weakness — to me as a leader. Author and professor Brené Brown states this perfectly in her book, *Daring Greatly,* when she says, "Vulnerability sounds like truth and feels like courage." Crisis arrives in a variety of forms, and its origins (as well as its solutions) are complex and tangled. It is impossible, therefore, for leaders to possess complete knowledge. Leaders are courageous when they confess their lack of knowledge. Those who attempt to fake their way through or become detached, irritated, or enraged are revealing cowardice, not courage. This behavior is not a sign of strength — it is a sign of an insecure and/or egotistical person. I've seen insecure leaders lash out in times of crisis, but I've also seen those who confessed their personal deficits. People gravitate toward the latter, not the former.

Leaders are courageous when they confess their lack of knowledge. Those who attempt to fake their way through or become detached, irritated, or enraged are revealing cowardice, not courage.

To understand vulnerability and transparency better, consider these examples. Imagine we all work at a software-as-a-service (SAAS) company and our server crashes (for more than just a few minutes), preventing the operation of our cornerstone software — leaving our clients in a lurch. A vulnerable leader would sound like this, "Well, this is a new phenomenon during my tenure. I've never seen this kind of catastrophic hardware failure without redundant equipment kicking in. Does anyone have any initial ideas on what steps can be taken to recover from this, and quickly?" Or "Teammates, you all are fully aware of the server crash that just occurred. My top three concerns are the loss of those customers who are less loyal to our service, the immediate impact to our cashflows, and what message we ought to broadcast to key stakeholders. I do have some thoughts, but don't have a solution at hand, so I'd like us all to share what steps the team ought to take to overcome." In both statements, the leader is exposing weakness and being brutally honest. In the second example, that courageous display of vulnerability and transparency will get subordinates thinking about and working toward a solution for or prevention of a major loss of customers, devastating impacts to cashflows, and what messages ought to be sent (when, how, and to whom). This is a powerful, not a weak, display of courage.

So, how do we hire or promote leaders with courageous characters? While courage builds naturally through experience, it is important to select senior leaders who already have courage, as described here, as a personality trait. Assessing courage in the first meeting with a person is challenging, but because it's so important, the attempt must be made. One way to do this is to offer fictitious crisis scenarios and ask the job candidate how they would react. You can ask them about their previous experience with crisis and how they or their leader reacted and what they learned from it. We live in an environment different from the days of surviving in the wild. Today's dangers come in the form of natural and man-made hazards which threaten the wellbeing of people, their livelihoods, and the survival of our businesses. This kind of danger requires a different form of courage as a foundational leadership ingredient in the crisis-ready organization.

Confidence

Confidence is the sibling of courage. Courageous leaders are confident leaders and vice versa. Confidence is demonstrated in several ways. It is demonstrated through the presence of a leader — how they carry themselves and express their demeanor — especially under duress. It is displayed in the knowledge that leaders possess about specific functions of a business, and it is found in the attitudes leaders exhibit when faced with daily challenges, big or small. The U.S. Army relies on the paradigm, *"Be, Know, and Do"* in its entry level leadership courses to develop future Army leaders.[18] Be, Know, and Do — by implying an expectation that leaders will always seek to understand fully, embody the mission and vision, and act appropriately and swiftly — captures the qualities a leader must possess to be confident and effective. Be, Know, and Do is an important ingredient in the organizational foundation recipe, especially as it applies to the trait of confidence.

Confidence is the sibling of courage.

Be equates to much of what I've included in the 5 C's — Calm, Communication, Courage, Confidence, and Creativity — the traits a leader must possess or work to develop. *Know* is both the knowledge of the tasks a leader must perform themselves and the tasks that the leader's subordinates must perform. The more a leader understands what their people do, the better they can lead. This said, I do not subscribe to "complete knowledge" as a prerequisite to leading. There was plenty I did not know as a leader, especially as the organizations I led became larger and larger. I could not operate a Howitzer artillery weapon myself or even as part of a team, but I knew its capabilities enough to lead the employment of it. Under the *Know* part of the paradigm, a leader ought to invest time in their personal development or seek help from others. When I was a young Army officer serving as the commander or CEO of an elite 50-man long-range surveillance detachment, I needed knowledge of what my teams did. One of our required skills was fast roping. Fast roping is a rapid and efficient, yet dangerous, method of tactical insertion for small teams into otherwise inaccessible areas like rooftops or tactically important terrain. In my CEO role, I would orchestrate the deployment of these

18 "Be, Know and Do," The Association of the United States Army, https://www.ausa.org/be-know-and-do.

teams. I wouldn't personally be fast roping onto a rooftop but I needed to understand how it was done to lead appropriately. As is customary in the Army, a noncommissioned officer (NCO) trained me how to fast rope. My senior NCO, an expert at fast-rope operations, took me aside personally and schooled me on the process. He watched me demonstrate it until I became proficient enough to understand the complexities of it. His instruction helped me gain confidence. I was confident in my abilities to perform the task with the proper instruction, but I needed an expert to show me. I was vulnerable and transparent enough to confess my lack of knowledge and experience and I was courageous enough to get "dirty" learning it. I was applying the *Know* part of the paradigm. The *Do* part of *Be, Know, and Do* relates to the act of leading. It includes acts like showing up early; engaging in self-development; being a champion of vision, values, and culture; and listening — to name only a few. Confident leaders lead regardless of the situation. Therefore, crisis requires confidence.

Confident leaders lead regardless of the situation. Therefore, crisis requires confidence.

As it turns out, it's not always easy to have confidence. Limiting beliefs are the biggest obstacle to confidence, and leaders must be on the lookout for them in themselves or in others. Asana, a business productivity organization, offers a fitting definition of limiting beliefs. "A limiting belief is a state of mind or belief about yourself (or your organization and/or the environment) that restricts you in some way."[19] In my coaching practice, I've seen limiting beliefs stifle growth and innovation. I have, of course, experienced limiting beliefs myself throughout my life and career. A limiting belief sounds like this: "I'm not experienced enough to handle this challenge." Limiting beliefs are countered with truths such as, "I have thought more about this challenge than others, and I'm a leader of character with good emotional intelligence — therefore, I'm well suited to meet and defeat

---

19 "10 Limiting Beliefs and How to Overcome Them," Asana, Inc., https://asana.com/resources/limiting-beliefs.

this challenge." When limiting beliefs creep in, as they often do, list out all the limiting beliefs on paper. Then, right next to them, write the truth. Following this exercise, you'll see just how futile the limiting beliefs were. Keep your limiting beliefs radar switched on!

There are leaders who are inherently confident in their abilities but, by and large, confidence is *learned* and *earned* along the way — as leaders bond with their people, forming trust and camaraderie. It is also born of the very things being presented in this chapter. An organization with a solid and resilient foundation will give leaders confidence when crisis hits. Emboldened by an organization with an inspiring cause, vision, and values — with a culture of inclusion, belonging and safety — a confident leader can emerge.

An organization with a solid and resilient foundation will give leaders confidence when crisis hits. Emboldened by an organization with an inspiring cause, vision, and values — with a culture of inclusion, belonging and safety — a confident leader can emerge.

Creativity

The last, but surely not least important, characteristic of our 5 C's is creativity. Creativity is more important today than it has ever been. The advent and emergence of technologies like blockchain, artificial intelligence, and advanced automation, robotics, and mechatronics have revolutionized modern business. What was once impossible is now possible. Creative leaders ask, "Why not?" to promote a spirit of creative problem-solving to achieve a solution that was once thought improbable. Those resistant to change will falter under crises, especially those crises that call for creative solutions and drastic change. An informative article titled "What is Creative Leadership?," authored by Menno van Dijk, Grant Davidson, and Valeria Mecozzi, offers this two-part definition of creativity, which I think is very useful as we examine the distinguishing traits, qualities, or properties of a crisis leader.

> "Creative leadership as a philosophy (or characteristic) embraces change as a given while seeking opportunity everywhere. It envisions desirable futures and unleashes the courage, collaboration, and creativity of contributors. Through a generous, inclusive purpose deeply rooted in pragmatic idealism and empathy, it gives rise to a transcendent consciousness that goes beyond individual gratification.
>
> "Creative leadership as an act builds on those desirable futures through scalable enterprises derived from innovative strategies. Creativity, critical analysis, experimentation, big vision, collaboration, bold action, calculated risk taking, agility and hard work all drive participative value..."[20]

When restaurants were restricted from seating people at 100% capacity during the height of the pandemic, revenue took an immediate hit. Effective leaders accepted and even embraced the change and sought opportunities and innovative strategies to overcome. They took bold action and led with agility toward a desirable future. Creativity flourished under this leadership. Restaurants with dining rooms converted to take-out and delivery operations. Wait staff took meals to cars in a parking lot instead of serving customers at tables. Hosts or hostesses took orders via the internet or over the phone instead taking reservations or seating people. And some restaurants built patios, rooftops, "igloos," and other outdoor seating options — knocking down walls and completely rethinking how to serve the masses in "socially distanced" ways.

Creativity, the fifth C, leverages the other C's of crisis leadership. Leaders display *courage* and *confidence* to foster creativity. They *communicate* a spirit of perseverance through creative thinking to their people. They *calmly* and positively receive creative ideas. Creative leaders strip off their blinders and attempt to do this with the people they lead. They bring

---

20 Van Dijk, M., Davidson, G., & Mecozzi, V. (2017, November). *Creative Leadership: School of Creative Leadership.* Retrieved from the School of Creative Leadership website. https://www.thnk.org/insights/what-is-creative-leadership/.

imagination into the culture of the organization where people can imagine a bigger, brighter future and take concrete steps toward achieving it.

Your stem-wall foundation is now set! Congratulations. Here is a crown made of laurels for your achievement (you victor, you!) but, as the saying goes, don't *rest* on your laurels. Your foundation will need some tending to. You'll need to clear brush and debris away. You'll need to paint it every few years and address the peeling and cracking, which occurs from weathering. In other words, you'll need to establish and promote your inspiring cause and vision every day. You'll need to work at getting your people to embrace the cause, vision, values, and culture of your organization. You'll need to attract, hire, and retain people who embrace and espouse these. If you or your people do not think realistically, crisis will exist only as a bad dream. Without effective leadership, no person will want to give crisis constant thought. What does deserve constant thought and nurturing, however, are the ingredients to the crisis-ready foundation: calm, confidence, communication, courage, and creativity. These foundational elements, added and cultivated under good leadership every day in your organization, will save the day when crisis hits.

You'll need to establish and promote your inspiring cause and vision every day. You'll need to work at getting your people to embrace the cause, vision, values, and culture of your organization. You'll need to attract, hire, and retain people who embrace and espouse these.

Following Devoe's death, my organization survived and thrived, mostly to honor his life, but also because we had built a strong and resilient, crisis-ready foundation. I will never quite shake the fact that Devoe never saw another sunrise over the desert after that IED exploded, but I know — without a shadow of a doubt — that more than 700 of us continued to wake up each morning, still mourning our comrade, but blessed to see the pink blossoms on the bushes and smell the naan bread baking in the ovens, not just because "life goes on" but because every single thing that we do "left and right of the boom" has potentially life-saving consequences. Because leadership matters, and never more so than before, during, and after crisis.

CHAPTER 4

Left of the Boom

Before Crisis Strikes

Now that we've explored how the crisis-ready foundation is set, let's invest more time considering what leaders should do in the relative peace of the "left-of-the-boom" (i.e., pre-crisis) space to prepare their organization for its next crisis. *Welcome to the left of the boom!*

It is here — "to the left" — where you and your business can take tangible steps toward mitigating the negative effects of crisis, before it happens (or before it happens again). Even as businesses move at break-neck speed amidst enormous workloads and myriad communications channels (e.g., email, web content, social media, ringing telephones), you can slow it all down with small, decisive actions that can prevent or mitigate their next crisis. In the Army, our left-of-the-boom space was filled by the training we did to prepare for war. Like crisis, war was imminent and the more work we did — the more priority we placed on preparation — the more likely we were to survive and win against a committed foe. Training and leader development were priorities in the units I served. Because of this, they were given focused planning, resources, and space on the calendar.

Of course, unlike the Army, not every business can carve out entire days or weeks for training (or they have, at least, not yet figured out how to do so without disrupting their operations or hurting their bottom line). At most small businesses and medium-sized businesses, the focus must be on the activities that directly bring it revenue. But what about the people and processes that *indirectly* enable revenue generation and, ultimately, profit? Too many of us have learned the hard way that if the proper emphasis is not placed on crisis preparation, revenue (or even the business as a whole) will suffer in the end. You can (and must!) dedicate an hour or two each week (or even just an hour or two each month) to left-of-the-boom activities. Here's how:

- You can weave in small chunks of time during existing events to imagine failures and talk about remedies; put this important work on the agenda for a team meeting, a leadership retreat, or a board meeting.
- Create stories of crisis and share them with your people — verbally, via email, or in your staff newsletter.
- Develop and disseminate short, digestible policy papers or standard procedures related to crisis prevention.
- Have your people read portions of this book and discuss it at meetings. (Invite me to your "book club," will ya? I'd be honored.)

Getting ready for a crisis is not a "nice to have" — it's a "must have" and you *must* find time for it. Whether you start with one of the tips above or launch your left-of-the-boom strategy in a different way, you must do *something* because it will save thousands, perhaps millions, of dollars. And it might even save lives.

You might be thinking, "But there are no body bags in my line of business, Rob. The stakes aren't that high." But the stakes are *always* high. In your business, human lives may not be at stake, but livelihoods and the life of the business is. And, as we have all seen, crisis often appears out of nowhere; therefore, your focus on left-of-the-boom activities will save the day when crisis arrives at your doorstep.

Getting Your Priorities Straight

It's hard to argue against this universal truth: One of the most vital responsibilities of a leader is the act of prioritization — of deciding what is most important and what deserves attention first, second, third, and onward … and of then fully and clearly communicating those organizational or team priorities to everyone in the business. Chief among those priorities should be looking at the horizon and planning for what's coming — not just what's on our desks or our "to-do" lists for today. And what's coming is often a crisis, a business disruption, a controversy, a scandal, or even a tragedy. It is the duty of a leader to prioritize left-of-the-boom activities so you and your people are as ready as you can possibly be.

It's hard to argue against this universal truth: One of the most vital responsibilities of a leader is the act of prioritization — of deciding what is most important and what deserves attention first, second, third, and onward … and of then fully and clearly communicating those organizational or team priorities to everyone in the business.

Priorities come in two forms: enduring and periodic. Enduring priorities are those most-important activities, which never change. Customer service is a good example of an enduring priority. Customer service always remains above everything else a business does. This never changes. Without satisfied, engaged, returning customers, you have nothing. Periodic priorities, on the other hand, come and go. They are important activities of a specific period. For instance, if a company was developing a new product, a periodic priority would be the project associated with the new product. All time and attention would be paid to it; alongside the focus on that periodic priority is your enduring priority, so the momentary priority would shift to the enduring priority of customer service if a customer walked through the door.

While carrying a hefty workload, leaders often struggle to create and enforce priorities. This is due to a legitimate fear that something on the "to-do" list will be ignored. Here's how I like to think about this: If priority #7 gets ignored, will our ship sink? If the answer is "no," than I have my priorities straight. If priority #1 or #2 gets ignored, it should mean that the organization and its people will take a significant hit. It could even mean the beginning of the end for the business. During busy times, there will always be work that gets lost on the back burner and it behooves you to ensure that such work is low-priority work so that you don't suffer a crisis as a result. There is no possible way a leader can delete all that an individual or an organization must do, putting the organization on autopilot and letting its employees sit back and rest. There will always be work, and while good leaders may be able to lessen the workload by focusing on what matters, eliminating waste, and achieving economies of scale, the amount of work your organization *could* and *should* do will still exceed the time and resources available. The only way to overcome the risk of working on the wrong things is through the disciplined work required to set, communicate, and enforce priorities.

Now that you have a philosophy or mindset for evaluating, forming, and enforcing priorities, try this exercise. Author and educator Stephen Covey offers his Time Management Matrix (or "4 Quadrants") to help leaders zero in on true priorities. There's nothing quite like forcing things into boxes or quadrants to smoke out the truth! The column headers (along the top) on the two-by-two matrix (resulting in four boxes or quadrants) are Urgent and Not Urgent. Along the left side from top to bottom, labeling the rows, are Important and Not Important. The real priorities for you and your team fall in the upper left quadrant: Urgent and Important, Quadrant 1.

	Urgent	**Not Urgent**
Important	QUADRANT 1 *-Do It-*	QUADRANT 2 *-Schedule It-*
Not Important	QUADRANT 3 *-Avoid It-*	QUADRANT 4 *-Delete It-*

Prioritizing your "to-dos" is easier when using Covey's Quadrants.

Using the Quadrants as a workplace exercise works best if you don't jump straight to filling in the matrix. Start *outside* Covey's quadrants by listing out everything that needs to be accomplished — your "to-do" list. Have a teammate or a small team create this list. I love a small team approach. Remember, stay outside the quadrants. Once you have your list, start moving things into the boxes or quadrants. Engage in short debates about what belongs where but get moving. Populate the 4 Quadrants. You can always move tasks but get something produced. Start with a draft and revisit it in a day or two if you must, but land on a decision or a firm list of priorities — urgent and important. You'll have done your job as a leader in the left-of-the-boom space.

Battlefield Tip: Use Sticky Notes for a Tactile "4 Quadrants" Experience

We live in a world with too many Word documents and Excel spreadsheets. If your team all work in the same location, you can organize your priorities first by doing so in a very tactile way, with the right office supplies. Try building your 4 Quadrants by having your team write each "to-do" item on a sticky note, such that everyone has several notes stuck to the conference room table in front of them. Then have everyone collect all their stickies, get up, and go to a dry erase board or a set of flip charts — one flip chart for each quadrant: Quadrant 1 (Urgent and Important — "do it!"), Quadrant 2 (Not Urgent, But Important - "schedule it!"), Quadrant 3 (Urgent, But Not Important — "avoid it!"), and Quadrant 4 (Not Urgent and Not Important — "delete it!"). Have the team members place the stickies on the appropriate quadrant.

In a matter of minutes, you'll be able to see — through a preponderance of sticky notes! — which types of activities are overwhelming your team. You might find that there are lots of activities that can be reduced or outsourced (Quadrant 3) or eliminated (Quadrant 4). Imagine saying to your team, "Let's not do that anymore"). Be sure to give ample attention to Quadrant 2 — these are your strategic issues that need to be scheduled and resourced appropriately, lest you find yourself in crisis by letting your foundation crumble. Quadrant 2 activities might not need to be done today, but they need to be done soon or eventually, and we can't lose track of them in the flurry of caring for Quadrant 1.

If "left-of-the-boom" activities are a priority, then you must prove it by having the discipline to schedule those activities and getting people to commit to doing their part. Fridays at 4:00 pm is a bad time for anything of importance. Approach left-of-the-boom events as company rituals. Rituals are activities that people commit to; they become common place to people and the organization and become "what we do" and "how we do it." Think about the kinds of rituals your family or your organization participates in — anything that you would never think of skipping. Maybe it's going to church each Sunday or brushing your teeth before bedtime, swiping your employee ID card to enter the building or showing up early for your boss's monthly sales-performance summit. In our organizations, left-of-the-boom activities could and should enjoy the same loyalty as whatever rituals are non-negotiable in your work and life. If you've ever been on a commercial airplane, you've heard the captain announce to the flight attendants that it's time to "complete pre-flight activities and secure the cabin." This is both a ritual and a left-of-the-boom, crisis-prevention activity.

REFLECTION QUESTION

Are your safety rituals and crisis-prevention activities effective or simply performative?

When it comes to choosing your left-of-the-boom, crisis-prevention activities and then ritualizing them, do you know how to hold yourself and your team accountable for doing it in a quality, meaningful way? Remember that "it's quality over quantity" and leaders must ensure these activities are productive and that outcomes are met. Simply committing to and conducting these activities is not enough. A ritual that doesn't strengthen you in some way is an empty ritual.

Six Steps to Crisis Readiness

In the spirit of keeping things simple (which is a sound operational approach), here is a "to-do" list for your left-of-the-boom space (not in order of importance). I'll walk you through each item in the following pages.

- ❑ Develop, train, and support a crisis-leader mindset — up, down, and across your organization
- ❑ Form crisis-management teams (CMTs) and give them their purposes, key tasks, and optimal end states
- ❑ Schedule time and space for teams to imagine, discuss and plan — regularly
- ❑ Provide an agenda (or the starting point for one) and collect outputs from CMT meetings
- ❑ Develop and brief a composite risk assessment
- ❑ Create the crisis-response plan

Crisis-Leader Mindset

As I shared earlier, a leader must honor the position they fill, not expect the position to bend or break to them. Leadership positions in an organization are created with the organization in mind, not the individual. These positions rarely, if ever, require a leader to get mired in the tactics of the moment. Leadership positions require the leader to think more strategically, analyze risk, and steer the organization. While it is completely appropriate for a leader to roll up their sleeves to solve a tactical problem or walk a subordinate through task completion, this ought to be the exception, not the norm. The proper leadership mindset fights the natural impulses of gravity — of being in the trenches with the troops instead of being up above, calling the shots, or being on deck with your sailors instead of being at the helm, captaining the ship. That impulse is natural and common — most of us rose to positions of leadership by first developing functional, tactical expertise; we know how to do the jobs that some of our people do, and we must resist the urge to keep doing it. We can only lead from above, like it or not. You'll need this mindset — the clarity to honor the position you fill — as you lead before, during, and after crisis ... left, in the middle of, and right of the boom.

Leadership can be a lonely business. While I'm a vulnerable leader, one who keeps others close and who shares my thinking and my doubts, I may be alone in considering how we could fail and/or scanning the horizon for threats. Even in small businesses — even in *family* businesses, where you're never truly "alone" and where people do really know and understand you — leadership is still a lonely affair, by design. Lonely as it may be, it is the job; to do the job, you need the mindset of a leader. Fortunately, it's a mindset that can be learned and honed over time. Subordinates, in their "busyness" (and you do the entire organization a disservice if you don't acknowledge that your people cope with a different state of being "busy" that's hectic in a way that your "busyness" is not) will complete tactical tasks contributing to the greater good of the organization but won't dedicate the time a leader would (or should) to contemplate the next crisis. A huge part of leading is thinking. And a huge part of leadership thinking involves *thinking ahead.*

A leadership mindset is a level of consciousness a leader maintains to best serve their position, the organization, and its people. It is a personal beacon that guides you in all that you do. Psychology educator and author Kendra Cherry describes mindset as "a set of beliefs that shape how you make sense of the world and yourself. It influences how you think, feel, and behave in any given situation."[21] You develop your mindset through introspection and determine which things are important to you — those values and priorities that you know will never change. When I took over as the commander (what the business world would call a CEO) of an Army brigade in the 101st Airborne Division, I focused less on establishing priorities and giving operational guidance — what exactly I *wanted* to do, on day one, but that I knew could wait. That guidance would come, but I focused instead on that which was *important* to me — for people to follow me, they needed to understand what I valued and what kind of mindset I was bringing to the unit. I focused on my mindset, as I knew this is how I would operate each day. My mindset was people-centric, and

---

21 Kendra Cherry, MSEd, "What Is a Mindset?" *Verywell Mind*, September 20, 2022, https://www.verywellmind.com/what-is-a-mindset-2795025.

I wanted everyone to understand that. A people-centric mindset means a leader will maintain a level of consciousness about the wellbeing of their people and how every decision affects them. People will be foremost in their mind in all that they do and in every decision they make. In the Army, I had the luxury of not being revenue-focused or profit-focused — our budgets would always be there, thanks to the federal government, and we weren't selling a product or service. I was able to focus on caring for my people and on being ready for the next crisis (of which there are many when you're an infantryman). In a nutshell, I was a people-centric leader with a crisis mindset.

A leader, when they adopt and embrace a crisis mindset, will dedicate time to ask (and answer) these questions:

1. "How will this fail?"
2. "What could be the next crisis for our team?"
3. "What signal is the environment, the economy, the government, or our competition sending?"

Philosophies to guide this mindset are:

- *"If I'm not dedicating time to imagining our next crisis, I'm doing a disservice to my people."*
- *"What am I doing — or not doing — that may bring about the next crisis?"*
- *"Do only those things that only I can do."*

That last philosophy takes us back to a leader's imperative to resist the urge to go into the trenches. To "do only the things that only you can do" means you will serve the organization as it must be served — that you will do what is required of the role. You will honor your position and delegate the rest. I realize that is easier said than done, especially when you're short-staffed, when your people are overwhelmed, or when you enjoy doing the "soldiering" work that you fell in love with long before you became a leader.

A great way to develop a crisis-leader mindset is to invest in a mirror. My mindset in crisis was always to hold up a mirror — to start with an analysis of *me*. When my organization failed or found itself in the throes of a crisis, I'd look at myself critically and consider what I did or failed to do that brought us to that moment. What did I fail to see? Did I not use the leader mindset described here? Was my head down in the tactics of the day as crisis loomed? I did this following the loss of Devoe. This mindset, contemplated internally and shared with others, will permeate the organization — readying it for battle.

As goes the leader, so goes the organization.

A great way to develop a crisis-leader mindset is to invest in a mirror. My mindset in crisis was always to hold up a mirror — to start with an analysis of me. When my organization failed or found itself in the throes of a crisis, I'd look at myself critically and consider what I did or failed to do that brought us to that moment.

Form Crisis-Management Teams

How do you know when it's time to form a crisis-management team (CMT)? Under the leader mindset, when you are contemplating threats to your organization, if one or more appear to have merit, it's time to form a CMT (or several of them). You know I love definitions, so let me offer a sound definition and approach to the forming and functioning of a CMT, which I've borrowed from the good folks at SmartSheet:

> *"A CMT prepares an organization to respond to potential emergencies (crises). It also executes and coordinates the response in the event of an actual disaster."*[22]

We manage *things* but we lead *people* and crises are things to be managed through the crisis spectrum of "left and right of the boom." Remember, we are still left of the boom, so it is analysis and planning that become our lifelines at this point.

The thoughts and inputs of a group or team are essential when it comes to imagining and planning for a crisis. In a CMT, it is necessary for people to make individual contributions, standing their ground with their input while engaging in respectful, yet constructive, debate. Avoid "group think" or assimilation culture, which is when individuals subordinate their thoughts to that of a group, choosing to think and behavior just like everyone else, despite their unique and meaningful perspectives.

To get started with CMTs at your organization, formalize your intent for the CMT(s). As I described in my first book (*It's Personal, Not Personnel*), intent consists of Purpose, Key Tasks, and End State. Here is an example intent for a CMT:

> CMT Orange (*give them a name*) exists to help prepare the organization for its next crisis *("why" or Purpose*). It will do this in Socratic style, leveraging the experience and diversity of team members imagining, debating, and assessing the impact of various crises to our organization.
>
> CMT Orange will produce the following (*Key Tasks*):
>
> 1. A comprehensive list of potential crises
> 2. Their impact to the organization
> 3. Steps to mitigate the effects of specific anticipated crises.
>
> When complete (*End State*), CMT Orange will produce a short report highlighting its findings and recommendations to company leadership.

---

22 Andy Marker, "How to Build an Effective Crisis Management Team," SmartSheet, May 13, 2020 (Updated October 3, 2022), https://www.smartsheet.com/content/crisis-management-team-roles.

You could be more specific, addressing a potential threat, such as a hack to a newly unveiled company software application. Some CMTs exist only for a certain timeframe and others are tasked with ongoing risk management; be clear in the CMT's charter about how long it will last and how or if it will be restarted if its work ceases at any point. For example, the CEO of a small trucking company in upstate New York might form a CMT for the duration of winter to analyze risks and the organization's vulnerabilities. The CEO might form this CMT in late summer to do some thinking and planning prior to the adverse weather season. It could be disbanded after the new year or shifted into a recall status if conditions changed. The charter of this CMT may come with some authorities for the CMT to form and implement safety procedures to mitigate risks, loss, and injuries. In developing the charter, the CEO could focus the CMT's scope on the operation of its fleet of trucks instead of a wide range of risks beyond trucking operations, such as parking lot safety and personally operated vehicle risks in adverse weather, to name a few. Finally, the CEO may empower the CMT to obtain resources and conduct training as needed.

Once you essentially have a "charter" for your CMT, it's time to assemble the players of the CMT. CMTs ought to be as diverse as possible in terms of background, experience, expertise, perspective, and other diversity criteria. Diversity works to broaden our knowledge and our decision-making power, and you must think of it far beyond skin color. Include the rank and file in these teams, if possible (not just people who are director-level and above — those same people who get assigned to every team, task force, or committee). Those who serve at the grassroots of our organizations have much to share. Be sure to gather people from as many disciplines or functions as you can, such as finance, legal, human resources, technology, manufacturing, logistics, customer support, sales, and marketing. Seek out military veterans (yes! for your civilian companies), as they are schooled at crisis preparation and response; they're calm under fire and tough as nails. Select people with experience brought forth from other organizations, from previous training, and from formal education. At its smallest, a CMT should be two people but might include a dozen or more players. Place someone in charge, let all players know who that person is, and empower them to lead the session and direct the people and the work. Lastly, consider outside expertise, such as consultants and/or trainers, to bring the needed level of experience to this space; sometimes the perspective you need doesn't exist in your organization just yet.

> There are a great many resources available online to help you think through the roles and responsibilities of the people on your CMTs, and I encourage you to check out some of the collective wisdom available to you if you're about to charter a CMT or two. In the spirit of keeping things simple, though, here I've created what I believe to be the core CMT structure required to help an organization in the left-of-the-boom space. You can build it out from here.
>
> **CMT Leader**. This person assembles the team and keeps discussions professional and constructive. They execute the agenda and form sub-teams as needed to accomplish meeting objectives. They are also responsible for meeting outputs. For this role, it is best to select someone with maturity and experience in previous crises. You may have to serve this position yourself in the left-of-the-boom space, especially if your organization is young, small, or hasn't had many crises yet.
>
> **Scribe**. This person captures meeting notes and assists with briefing production. They conduct a review of previous meetings, including tasks to individuals and deadlines. If there are only two members in the CMT, a crisis team member can also serve as the scribe.
>
> **Crisis Team Member**. Every time member is arrive prepared to imagine crisis scenarios, identify their impacts to the organization, and develop measures to mitigate damages. Every member matters.

Rule #1 in a crisis-management team is that you can't manage anything if you don't make the time. As such, you must schedule time and space for teams to imagine, discuss, and plan. Under your intent and established priorities, schedule time and space for teams to do the work required to mitigate the boom. Help them *make* space by taking something off their plates or cancelling a nonessential activity, meeting, or event and then telling your employees that "This 'found time' next Tuesday morning would be well spent by gathering your CMT. I know you all have time now." With some preparation of the CMT Leader, 1-2 hours could be sufficient to achieve a minimum level of crisis analysis and planning, but keep it boxed in with a specific duration to prevent a meeting that drags on. These meetings ought to be held away from the office, if possible, in a distraction-free space where people can think, talk, and share ideas. Cellular phones and other distractions ought to be kept from CMT meetings, if possible.

For that first CMT meeting, don't expect to accomplish much if you start from a blank piece of paper. Begin with the end in mind and start with a clear agenda. Whether you are the CMT leader or the business president or other leader responsible for organizing these teams, provide an agenda and collect outputs from CMT meetings (or hold a trusted person responsible for these key activities).

Fair warning: CMT meetings will spin out of control if not directed properly. Time spent by leaders giving intent, establishing priorities, selecting the right people, and defining critical concepts like "crisis" and "decisive" will determine their success. Again, keep these meetings simple and focused. If you are forming a CMT to discover crises not yet imagined, you'll want to carve out more time or consider a series of meetings to flesh out as many possibilities as you can. Invest time with the CMT leader — sharing your experience, talking over your intent, and articulating what concerns you have as a leader. Help them understand what you mean when you say "crisis" and "decisive." Have them read Chapter 1 of this book. Focus on those potential crises that may set the organization on a path to strategic failure — on a potential path to its ultimate demise. With the leader, sketch out an agenda, rehearse how the meeting might go, then trust them enough to let them run with it. Reinforce "imagination." It will be the role of the CMT to imagine the unimaginable and they ought to feel free to do so. If you have specific potential crises in mind, share those and help define them as clearly as you can. Avoid giving solutions. Let the CMT come to you with those. Instead, invest your time in understanding and defining "crisis" in a broader way to guide the team.

A precise agenda for a crisis-management team (CMT) meeting could look like this:

1. **Define the crisis.** Name it or them and describe how it is a crisis — how it might mean the end of the organization. Keep it short. A sentence or two would be best.

2. **Build the narrative.** This is where the CMT can create a fictitious story that describes how the crisis came about and what its effects are to the organization. The story could end with the ultimate demise of the business. Don't sugar-coat it but don't be wildly unrealistic either. Under the leadership of the CMT Leader, tell a realistic (but potentially terrifying) story of the crisis.

3. **Develop solutions or mitigating activities.** This may be the hardest part of the meeting, but it must be an agenda item. Discuss and develop something — even if there is a chance it might not work. Some crises may not be solvable but mitigating activities can lessen their impact, even allowing the organization some time and space to counter their negative effects.

So that you can see how this all comes together in practice, let me provide an example of a CMT — a fictious crisis-management team at a fictitious company, drawn from the realities of the small businesses that I own. Hello, I'm John (sort of).

John owns and operates a painting company that offers all-surface painting for residential and commercial customers. He buys his paint from ABC Paint Supply, who informs him that in two months' time, orders for paint will experience delays due to a variety of supply-chain and factory-staffing issues. John should expect shipping delays and out-of-stock notices during the next nine months or more as ABC completes a new production facility and executes its hiring campaign. John has a crisis-leader mindset and is always scanning the horizon for threats. He knows there are no assurances of business survival in today's challenging and changing environment. He has several competitors who offer paint from a variety of sources, though not of the best quality. Knowing his current and future customers will want the same timely and quality service, John decides to form a CMT to analyze this challenge and offer solutions before the product shortage begins to impact his business. John knows he is in the left-of-the-boom space. Nothing has "blown up" quite yet, but he can hear the rumblings.

John selects his most experienced and respected salesperson, Sarah, to take the lead. He knows Sarah has the experience with the industry and market (i.e., she has the necessary knowledge) and the demeanor and leadership acumen (i.e., she has the necessary skills and attitudes) to lead a CMT. He gives Sarah two other people who have some experience and who have shown commitment to the company and its future. He offers this intent to Sarah"

> "Sarah, the *purpose* of this crisis-management team — this CMT — is to prevent business failure given the recent news from ABC Paint Supply. Here is Rob Campbell's book on crisis. I'd like you to read Chapter 1 at a minimum before you form and lead the CMT. Here are some *key tasks*. 1. Contact ABC Paint Supply to gather more fidelity on their challenges and the ramifications to our business. 2. Define the crisis. 3. Create a 'narrative of failure' associated with each crisis you define. 4. Offer a solution or solutions to each crisis. Focus on a handful of threats as imagined by the group. We can expand later as desired. The *end state* for the CMT will involve you holding a short but informative, well-prepared briefing that helps me understand all the risks and helps me determine which decisive steps I can take to help us overcome the challenges stemming from this pending issue."

Sarah calls the CMT together and spends a preponderance of the time defining the competitive environment — other painting companies, what they offer, customer trends, and other characteristics. She focuses the group on identifying problems and their causes and impacts to the business. She keeps the team from rushing to solutions — a natural tendency of individuals and groups. John checks in only to see if more guidance is needed or if there are any questions he might be able to answer. He avoids getting into the weeds or stifling creativity. Once a foundation of environmental and problem understanding is set, Sarah steers the team toward developing solutions. She allows the CMT to think and converse without boundaries, meaning that they are free to imagine creative solutions and their likely impacts. No idea is shirked. Sarah serves as the Scribe and spends time on a dry erase board, capturing ideas then, with the group's help, narrowing them down to three main suggestions to present to John.

Keep the crisis-management team (CMT) from rushing to solutions – a natural tendency of individuals and groups.

The CMT then develops a short but informative brief in which they meet John's intent. The brief can be written but it is best delivered verbally to allow for dialogue and feedback. The brief expands on the challenges facing ABC Paint Supply and their impacts on John's business. The brief then defines the crisis, starting with a description of the current environment, with particular focus on industry and customer demands. The CMT describes three crises: one centered on supply-chain challenges, another involving a shift in customer loyalty, and a third resulting in a reduction in quality for which the business is known. Each of these — or a combination of them — could result in catastrophic failure for the business. Each of the crisis scenarios is presented in a story format and each is capped off with suggestions to mitigate negative impacts. Suggestions include a series of reassuring strategic messages the business can send to its customers, a broadening of vendors to compensate for supply-chain challenges, and an expanded quality-assurance and quality-control approach the business can implement. The outputs of this CMT brief will allow John and his team to take the next step, which is the development of a risk assessment and crisis-action plan and, of course, John will share the CMT's brief with the greater team.

Risk Assessment

The risk assessment is a separate and distinct document. It may or may not be the product of a CMT but it certainly should be informed by it.

Whether a CMT brief is done or not, a risk assessment should be done and documented for leaders to examine and to guide their decisions. "Seeing" risk through a formal document helps a leader and their team prevent damage, injury, or failure. The CMT's work helps populate the risk assessment, either immediately or shortly thereafter.

The most important thing to know about a risk assessment is that it is only a piece of paper — a written assessment that, if not acted upon, is worthless to an organization. I recall this lesson vividly from my Army training, when we pored over risk assessments and when I first asked myself, "How will we actually conduct that step to mitigate risk? Who, what, where, when, and how will that specific mitigator be implemented?" The assessment is the *start* and an important one at that.

The U.S. Army has an entire manual dedicated to risk management, *Army Pamphlet 385-30*. Here I get right to the core of it. There are five areas of consideration in analyzing and mitigating risk:

1. Hazard identification
2. Initial level of risk
3. Controls or mitigating steps
4. How and who will implement
5. Residual risk level.

Going back to our company software example from Chapters 1 and 3, here is what a risk assessment would look like.

1. **Hazard identification:** Potential for someone to hack the new application (The more specific the definition of the hazard the better).
2. **Initial level of risk:** A number can be assigned as the initial level of risk. With a scale of 5 being the highest (most risky) and 0 being the lowest (least risky), let's say that this hazard rates as a 4 in its level of risk to the organization.
3. **Controls or mitigating steps:** A control or mitigating step (and there can be several) could be a detection application

(or anti-malware) built into the software to monitor nefarious activity and detect hacking. A control can also be a person tasked to identify signs specific to this risk — as early as possible — to prevent their impact.

4. **How and who will implement:** Implementation could be assigned to the information technology (IT) director or another capable person with resources and authority to implement the controls or mitigating steps. Here, a description of how this mitigation would be accomplished will force the person creating this report to apply a dose of reality toward a mitigating step or solution. "How" could be summarized as the steps taken to install the anti-malware application and/or ways nefarious activity will be detected and stopped. "How" can also be quick or more deliberate inspections on a recurring basis.

5. **Residual risk level:** In this case (as in many), it may be hard to mitigate the risk completely. But by going through the steps of risk mitigation, the risk level could drop to a 2 or 1, to make it less risky for the organization.

When it comes to doing a risk assessment, I suggest you keep it simple but take the time to "fill in the blanks." In so doing, you'll have gone through the process of managing risk, a vitally important left-of-the-boom activity. Few things are fail-safe but by going through this process, the chances of catastrophic failure decrease and that may be the very thing that keeps your organization afloat. Lastly, be cautious of determining (and celebrating) that there's zero risk. It may look comforting in a presentation, but some risk will almost always exist. We are at risk in our own living rooms. Arriving at zero risk means there are no mitigators required. Even in our own living rooms, the positioning of the furniture, the wiring of electronics, and the traction beneath our feet (rugs and/or floors) are areas of risk requiring assessment and mitigation, albeit less formal. You don't keep the living-room windows open during a thunderstorm and you don't let the kids leave Legos all over the floor — risk exists everywhere and risk mitigators are our constant friends.

Here is an example of a risk assessment. I've taken this from the actual *U.S. Army Composite Risk Management Worksheet* and simplified it for you. The act of creating an assessment like this is an excellent left-of-the-boom activity. Let me walk you through it:

- Column 1 describes the **hazard**. Be clear on this. Keep it to three words or less.
- In Column 2, the **initial level of risk** should be assessed as if no (or few) mitigators were in place. Even in a car with seatbelts and airbags, there can be greater risk than if further measures were implemented (e.g., reduced speed, proper lighting, safety cameras).
- Column 3 is for **Controls** — those actions you'll take to mitigate the risk. Get creative here, then get logical. My caution to you in this block is to create controls that will actually mitigate the risk yet are realistic. For instance, it is not realistic to assign a person to monitor something for several hours straight. The controls column is the most important one. Here is where leaders and teams do the hard disciplined work to "lessen the boom."
- Column 4 is where you enter the risk that exists *following* the mitigating steps you'll take: the "**residual level of risk**." Note that I've only given the options of High, Medium, and Low. There is always some risk.
- Column 5 should not replicate Column 3 but is for checklists and rehearsals where responsible individuals run scenarios and check off action steps for Column 3's controls. This column is about **implementation**.
- Column 6 is the person responsible for overseeing the risk mitigation. Indeed, the leader is overall responsible, but this is the leader's and the organization's representative — a single person who owns this part of the risk assessment. This last piece of the risk assessment is about **how to supervise** the mitigation and potential response to the risk. You can create a table like this, sketch it out on a dry erase board, or even think and talk through this as an individual or a team. Whatever you choose, you should always choose to assess risk formally.

Mission/Task:		Start Date:		End Date:	
Prepared By:				Date Prepared:	
1: Hazard	2: Initial Level of Risk (L/M/H)	3: Controls	4: Residual Level of Risk (L/M/H)	5: How to Implement	6: How to Supervise (Who)
Icy Roads	High	• Vehicle tire chains • Delay operations until 11:00 am	Medium	• Driver notification roster and rehearsal prior to closing • Vehicle inspection evening prior	Fleet Manager
Theft of Grand Re-Opening Signage	Medium	• Spotlights left on • Police notified • Web-based cameras installed • Signage chained to concrete	Low	• Final inspection prior to closing	Facility Manager
Server Crash	High	• XYZ IT server monitor • Morning & Evening diagnosis • Back up server	Medium	• Opening and closing diagnostics check • Weekly XYZ server health report	IT Manager

Source: *United States Army Field Manual 5-19*, "Risk Management."

Create the Crisis-Response Plan

Now that you may have formed a CMT and you have a risk assessment, it's time to create a short and impactful crisis-response plan. I suggest you use this format:

Spell out *who* does *what* to *whom* — *when, where, how,* and *why.*

For this crisis-response plan, I will stay with our company software example. Should our company's servers be hacked, what would the company do? The more questions you can answer here, the closer you'll be to having a response plan. Start rough, sketching the response out on a dry-erase board or blank sheet of paper (or a virtual white board or shared online document). Start with the event: "We were just hacked." Then start writing out the 5 Ws (who, what, where, when, why, and how).[23] In our example, once hacked, there would be a reporting requirement to the person who could mitigate the damage best, then to select leaders in the company. There might be a company-wide notification, including immediate actions to be taken by individuals. There might be some immediate steps to prevent further damage, such as turning off servers, disconnecting computers from networks, etc. Note that many of these events may happen (and need to happen) simultaneously. Customers and other stakeholders may be notified at some point in the process and backup systems initiated to help recover from damage (and recover missing data). There may be financial implications and actions, such as minimizing expenses and even shrinking payroll (e.g., halting bonuses, stopping overtime, even terminating employees), if needed, for survival. Dates and times could be included to give all involved a sense of timing "post-boom." For instance, "Hack (or insert crisis name) + 10 minutes" might be the goal for disconnecting all necessary automation. "Hack + 4 hours" might be the goal for informing customers. From this exercise, additional tasks will sprout, further mitigating the damage or preventing future damage. Drafts of public relations (PR) messages, additional protective software, and a company employee notification roster or phone tree may materialize.

---

23 The 5 Ws model actually contains six elements: five Ws and one H.

Battlefield Tip: Have Your Crisis "Cheat Sheets" Ready and on Display

At work and in life, there are many types of crises or emergencies that might occur, and while you can't be fully ready for them all, you *can* be ready for the most predictable ones. And you *must* be ready for the life-and-death ones.

Someone once told me a terrifying story with a happy ending — a happy ending that was made possible because a bank's leadership team (maybe because of a crisis-management team!) had the foresight to post "In Case of Emergency" documents in every cubicle, office, and common space of their multi-story building. Those simple pieces of paper indicated who — on each floor of the building — had first aid training and AED (automated external defibrillator) training. So, on a very scary day, when someone in the mortgage underwriting department noticed their coworker slumped over her desk, they were able to consultant the "cheat sheet" and call their floor's cardiac-crisis "point person," who immediately came running and began performing cardiopulmonary resuscitation (CPR). The bank employee who had a heart attack survived. And the paramedics who eventually took over CPR were clear that if not for that "first-aid cheat sheet," the crisis surely would have ended in tragedy.

Not every workplace is a battlefield — in fact, most of them are not. But every workplace is ripe for crisis and just waiting to test its people for how well they prepared when they were "left of the boom." When I was leading my unit in Afghanistan, we spent most of our time "left of the boom" — doing everything I describe in this chapter and doing it long before Devoe was killed. Indeed, training for crisis — preparing for combat — was our principal focus. This doesn't mean that things always went smoothly when crisis hit us and this fact underscores the importance of "left-of-the-boom" activities; if well-trained soldiers sometimes stumble during and after a firefight, so will you. Amidst a demanding schedule with new team members arriving almost daily, we stuck to our war-fighting — with "survive and win" priorities — yet through all our training, we inserted scenarios where individuals and leaders were injured or killed or where things went wrong. Unlike some non-military organizations, we never pretended that things would never go wrong, and we never got so wrapped up in what was on our proverbial desks (our short-term work) that we forgot about training and planning (our long-term imperative). This left-of-the-boom focus paid off when faced with crisis on the battlefield.

If well-trained soldiers sometimes stumble during and after a firefight, so will you.

In training, throughout my 27-year Army career, leaders across the organization asked the hard questions, like "How will this fail?" Crisis-management teams were formed often to imagine the unimaginable and develop preventive and reactionary activities. We spent time thinking, analyzing, and drilling to flesh out mitigating steps we could take to not only *prevent* crisis (when possible) but *mitigate* its damaging effects (when crisis was unavoidable). And we did this extensively as we prepared for our renewed mission in Afghanistan in 2009, when my unit was among the additional 33,000 troops sent to the region. We would be new to Afghanistan and there was much we would learn. One of our primary goals was to manage crisis, not let it manage us.

We were not perfect. Had we been, Devoe and others might still be alive. But because we did what we did in our left-of-the-boom space, I submit, others did survive. We, as a team, were able to survive a horrific event on that early March 2009 morning. We lost a brother and it put us on our heels, but the unit and its people survived to fight another day.

CHAPTER 5

The Boom!

During the Crisis

BOOM! An explosion happens, your company computer system crashes, a power outage, you just lost your biggest client, a pandemic or a storm hits and you are directed to close your business, or God forbid, there is a serious injury or fatality on your team. Your blood runs cold. You are momentarily paralyzed. You might gasp, shout an explicative, or run for the cover of a safer space. Seconds become minutes, and minutes become hours — all the motion around you halts, and you're quite sure that the whole world has stopped turning on its axis. At once, the images, words, smells, and sounds, are all seared in your memory forever. Crisis has struck.

We all react differently to a boom — to a crisis, emergency, tragedy, disruption, scandal, or controversy — and our responses, varied as they may be, are natural and valid. I still remember these feelings and this sense of time slowing down or stopping when I learned of Devoe's death, when I experienced the death of other soldiers and friends of mine, and when under enemy fire during my Army career. Most booms last only briefly and what follows is the ultimate test of leadership. Star NFL quarterback Tom Brady said, "A great quarterback is one who can perform under pressure." Very true and there is no greater pressure — no greater need for effective leadership — than during a crisis or in the moments immediately following it. When managing a crisis, we are all "quarter-backs" — making critical decisions about where to throw the ball next, hoping we get it right and hoping our muscle memory doesn't let us down. It is here — during the boom, when the smoke is still rising and our ears are still ringing — where leadership succeeds or fails. The wrong reaction, including no reaction at all, can have a significant and lasting impact on your organization and your stakeholders. Do something. Don't freeze. Your people will look to you upon crisis; they might freeze but you cannot. So, so many people are counting on you.

Most booms last only briefly and what follows is the ultimate test of leadership.

The Most Important Time to Lead

Recall the 5 Cs of Crisis Leadership — Calm, Communication, Courage, Confidence, and Creativity — highlighted in Chapter 3, "Establishing a Crisis-Ready Foundation." It's time, during the boom. to put those into action. John C. Maxwell, leadership speaker and author says, "Leadership is influence — nothing more, nothing less." Leadership during a crisis puts those influencing talents to the test; crisis is a peak time to influence others by your words and your example. You should be the first to step forward, to act, and to speak up when crisis hits. Be calm but lead. It is vital that *you* lead the crisis, and that you don't let *it* lead you. After a boom, all eyes and ears turn to leadership. Give them something to look at, something to hear, and something to believe in.

My leadership approach has always been to remain calm, to ask pointed questions to fill in my information gaps, and to offer assurance that we, as a team, will persevere. I will likely direct immediate actions to minimize impact and prevent injury and further damage. I may simultaneously seek input, displaying (I hope) vulnerability and humility, saying, "OK, I'm open to input and ideas as we navigate this challenge." At the moment of the boom, I may act faster and more decisive — receiving input, processing it, accepting or discarding some, and making decisions as quickly and as accurately as I can. Bold actions and decision-making may take the place of my calm, deliberate leadership approach, but I'll always do my best to be assured and decisive, even if observers might not describe my crisis responses to be "calm." I'll let the situation dictate whether I get busy listening or get busy issuing instructions (or a little bit of both).

Muster Up Some Courage

Poet, author, and civil rights activist Maya Angelou once said, "Courage is the most important of all the virtues, because without courage, you can't practice any other virtue consistently." And she was right. You'll need to be courageous when things go "boom!" There is no other option. I've seen countless leaders act courageously in the direst situations. Each of them, if they are honest, will tell you they were scared to death but knew they had to act — so they "saddled up" and rode on. In the left-of-the-boom space, you should invest time studying fear and courage, and imagining possible (and impossible) scenarios and thinking about how you will act. Maggie Wooll, an expert on the evolving future of work, teaches us that it's possible to *learn* to be brave, and that there are nine ways to be courageous. Her nine steps to courage include:

1. Acknowledge your fear
2. Confront it
3. Use positive self-talk
4. Get comfortable with being uncomfortable
5. Make being brave a habit
6. Practice mindfulness
7. Find a mentor
8. Embrace failure
9. Don't aim for perfection.[24]

---

24 Maggie Wooll, MBA, "Be Brave Even If You Aren't Yet (9 Ways to Conquer Your Fears)," *BetterUp*, July 22, 2021, https://www.betterup.com/blog/how-to-be-brave.

These are all sound actions a leader can take, even in the "during-the-crisis" space, specifically #1, #2, and #3. Even in the fleeting seconds of a boom, a leader can acknowledge and confront their fear using positive self-talk. Inside the head of a leader, it might sound something like this, "Ok self, this just hit me like a punch in the face. I'm scared." Or "I acknowledge this fear, but I know I'm a good person and an effective leader. My people are with me. I'm not alone. Saddle up, self! Lead!" This might sound a little hokey, but even the most courageous leaders do exactly this. I did it upon learning of the death of Devoe. I don't remember my exact words, but I know I spoke to myself as I moved toward the operations center to receive what I knew was very bad news. I'm sure my internal self, delivering a silent monologue, reminded me to be calm and think. I know I challenged myself with something like, "OK, Campbell ... saddle up and lead."

Remember that activities in the left-of-the-boom space will contribute greatly to the courage you must display when the boom hits. It must be restated — proper preparation enables critical courage. Creating and playing out scenarios and imagining or acting out how you would respond will serve as mental models when crisis arrives. Countless times in combat, we succeeded because we had drilled ourselves hard and repeatedly. Every little bit helps. Huddle your team often and work through scenarios. Use this book. You'll be more courageous for it in the end.

Remember that activities in the left-of-the-boom space will contribute greatly to the courage you must display when the boom hits. It must be restated — proper preparation enables critical courage.

Set the Tone, Communicate, and Act

A leader's #1 job during a crisis is to remain calm. When crisis happens, people look upward. Whether the situation involves a local or national government during a pandemic or a company that just lost its largest customer, people will look upward toward leadership for answers and reassurance. Leaders ought to give them something to look at: calm, effective leadership. I often say, "Leaders set tone." I usually repeat it when I say it, to emphasize its importance. "Leaders set tone. *Leaders* set *tone*." Organizations inevitably take on the demeanor of their leader, whether good or bad. If the leader is toxic, the organization will be in fight-or-flight mode and the environment will be extremely negative. Toxicity may even be emulated within the rank-and-file. If a leader is caring, enthusiastic, charismatic, and vulnerable, the organization will imitate this, and people will gravitate toward these leadership characteristics. The reaction of people and of the environment — whether under good or bad leadership — will only intensify when crisis arrives. Crisis is a magnifying glass.

If a leader is caring, enthusiastic, charismatic, and vulnerable, the organization will imitate this, and people will gravitate toward these leadership characteristics. The reaction of people and of the environment — whether under good or bad leadership — will only intensify when crisis arrives. Crisis is a magnifying glass.

When I learned of Devoe's death, I could have lost control. I could have immediately questioned leaders and demanded answers for how this could have happened. I could have become personally emotional at the loss of human life. I could have gone draconian, bringing immediate and sweeping change to the organization in terms of how we operated. I could have shut down. To do all of these things would have been human. But I knew better. I had witnessed crisis up close and had learned from the best. I remained calm and asked pointed questions — emboldened by trust — that helped to fill in my information gaps and guide my actions and decisions. I was quiet and focused, though my mind was racing.

While calm, I did let my emotions creep up on me. I was separated from my troops — my Army brothers and sisters — and I wanted to be by their side in this time of grief. I felt powerless and wanted, in some way, to regain that power. As I admitted in Chapter 3, I reached out to the aviation battalion commander, a peer, to see if he could provide a helicopter for me to return to my base a half-day's drive away, over very treacherous and enemy-infested terrain. I was *reacting* more than *thinking* deliberately but, with his help, I regained my composure. He was very sympathetic and understanding, but he rationalized why it would be unwise to provide a helicopter to me at that moment. There were other critical missions happening, which were part of a greater campaign to achieve victory, and my stripping away a helicopter would have disrupted the greater plan. His voice of reason was what I needed. I could quickly see that my need was subordinate to the greater need for lift aviation. For all the preparation I had done in the "left-of-the-boom" space, I still needed to the support and calm reasoning of someone else to execute my crisis response wisely when I was truly in it — when the "boom" was under way and it was time to prove my mettle when it comes to crisis response. That aviation battalion commander, perhaps not as overcome by emotion (shock, grief) as I was, was able to put me back on track in a matter of seconds. Once I had a moment to process his input, I was able to step back and see the big picture immediately. I knew that I had good leaders — people I trusted — on the ground. They would do the right thing and would carry the day until my return. I knew there was nothing I could do for Devoe anymore. As it was, his remains would be flown to the base I was currently on, and I would be there to honor him and to care for a few of my soldiers who were collocated. I would remain where I was, communicate clearly, set the tone, and act.

While I strongly advocate remaining calm when things go "Boom!," there are moments when a leader must step forward and direct action with courage and emotion. These moments are usually when lives or livelihoods are at stake. If a computer virus hit a SAAS company, it would be completely normal for a leader to run from person to person, desk to desk, directing people to unplug their computers. Time would be of the essence and the impact of the virus great if a leader did not act decisively and with emotion. I do submit these times are rare. Usually, even in the direst moments, it is a leader's job to remain calm. This tone of calm focus is displayed by how a leader communicates. Recall the over-dramatic news reporting during hurricanes I previously mentioned. Apocalyptic words and actions will emit a response — but not the one you want.

Even in the direst moments, it is a leader's job to remain calm. This tone of calm focus is displayed by how a leader communicates. Recall the over-dramatic news reporting during hurricanes I previously mentioned. Apocalyptic words and actions will emit a response — but not the one you want.

Communicate, Communicate, Communicate

The first two questions we inevitably ask ourselves and each other when crisis strikes are "What is happening?" and "Who else needs to know?" Crisis is marked by ambiguity and by events happening at rapid speed. Once the leader has control of their emotions and has set a tone of focus and calm, it's time to overcommunicate. Crisis calls for frequent communications. Quite often, on the right side of the boom — when errors are discovered, and lessons are gathered — a lack of communication is found to have contributed to failures. Overcommunicating does not mean communicating for the sake of communicating: sending constant, stream-of-consciousness texts and emails or picking up the phone every few minutes. It does mean that everyone (leaders as well as members of the whole team) should share information frequently — more than they normally would — as a situation develops. In crisis, the need for information increases.

To guide you and especially your team, ask and answer this two-part question often: "What is happening and who else needs to know?" We used this in combat to help others around us react and support the greater mission. People at all levels can help solve problems but only if they are aware of what is happening. In crisis, information will help assuage the anguish associated with the event. When leaders go silent, angst increases followed by fight or flight responses. The organization risks failure. Moreover, information will help leaders and subordinates react in a more informed manner and make better decisions.

REFLECTION QUESTION

More information, please! How do you perform when it comes to sharing enough information during a crisis?

When has receiving or giving extra information made you or others feel better during tough times? Think about a time when someone had to deliver bad news to you — maybe even tragic, devastating news. If you're like most people, there have been times like this when getting as much information as is reasonable (as much as is *possible* is sometimes too much information) has left you saying, "Thank you so much for letting me know." There is gratitude in the simple act of receiving information — of being brought into the loop when things feel out of control. When something goes "Boom!" and if it impacts us personally or professionally, we want to know — even if there's nothing we can do about it. Recognizing that, how might you think about the concept of "need-to-know" information differently the next time crisis strikes?

To communicate optimally when crisis strikes, we need to have prepared adequately in advance. Again, leverage your left-of-the-boom activities and the sort of "muscle memory" of CMT conversations and briefs, drills, training, and tabletop scenarios. Under the chaos of the boom, it is possible to forget the work and preparation you did. Effective leaders turn back to that space, even if only briefly, to see if they can draw upon its lessons. Revisit scenarios you may have played out. Review risk assessments and their mitigating steps. Form CMTs again, as needed. You'll find your left-of-the-boom work will pay off in the madness of crisis.

As they say, timing is everything. And crisis always happens at the worst time — in the middle of the night, on a Friday afternoon, in the early morning, or on the weekend. Combat was famous for that. The enemy (crisis) got a vote and they typically voted when they knew we would be at our most fatigued, unable to think and lead effectively. To guide

Army leaders in succinct, useful communication, we used easy-to-follow formats, such as the Situation Report (SitRep or SITREP). In a joking (but effective) way, I use it with my friends and family daily. I may text "SITREP" — as I did in the Army via secure radio or voice over internet protocol (VoIP) communications — to see what friends and family are up to or how they are doing. In training and combat, it sounded as simple as this: "Taskmaster 6 [a commander's tactical callsign], this is Bastogne 6 [my callsign]. SITREP, over." The commander would know what to do. He or she would send me an update on their situation. It might sound something like this: "Bastogne 6, this is Taskmaster 6, I've got four support units in position to refuel and restock battalions on their separate routes to the main objective. The quick reaction force is 30 minutes from being fully ready. I'll report when they are set." This took some training, and you can absolutely practice it in your organization, regardless of your industry. Use "SITREP" as a company acronym but train your leaders and people what you are looking for. Use a simple format, easy for them to remember when tired and shocked by the boom, such as the 5 Ws — Who, What, Where, When, Why, and How. Have them fill in those blanks to the best of their ability. They will get better through repetition.

Here is an example of how to use situation reports in business environments. Andy knows he had three customers this week who were slightly displeased with their service. While he trusts his team, especially his service manager Mark, it is his duty to check on these relationships. They are vital to his business's success. Andy sends Mark a text message, "Hey man, send a SITREP on our disgruntled customers by end of the day, please." Mark responds using a 5 Ws format, describing the situation with each customer:

> "Mrs. Teal [*who*] understands the damaged product was no fault of ours [*what*]. Our distributor is using a different shipping company as of last Monday [*why*]. Mrs. Teal is disappointed in the delay [*what*] as she had plans but I had a chance to talk to her again this morning [*when*] and assure her we would see this project through. I'll reconnect with her on Monday, when we should have a new ship date [*when*]. Marjorie is standing by to schedule the team to return to Mrs. Teal's home once we receive the new product [*what, how*]."

This SITREP is quick but thorough. It allows Andy to make decisions if needed. It helps him lead, especially in crisis.

Communication during the boom comes in two general forms. The first is the initial report. The second is follow-up reporting. All reporting is important, but especially the initial report — as it will guide the first actions of a leader and an organization. The initial report should, at a minimum, follow a format like the one below. This sample scenario involves the loss of a client.

> "Matt, tell me what just happened!"
>
> Matt uses a 4-step initial report: 1. What happened. 2. What actions I took or am taking. 3. Who knows. 4. What I recommend as next steps.
>
> "Boss, client X just called and canceled all future services [*what happened*]. They did not say why but Morgan is on the phone with them trying to figure this out [*what is being done*]. I've notified the marketing director, vice president of sales, and finance [*what actions I took*]. I suggest we form a CMT team to review impacts and talk about initial steps we can take [*what I recommend as next steps*].

Matt's follow-up report generally follows the same 4-step format, but this time he has more information to offer.

> "Boss, Morgan's attempt to salvage this account has failed. She sent an update in the company chat application to all involved, including you. Sales and finance have offered their initial thoughts on the impact of losing the account. Here they are. The CMT team is set to meet in one hour. We may ask for you to engage with the client directly, but I want to provide you with some analysis and recommendations before you do. The CMT team will flesh that out and report to you following its huddle."

In between these reports, in the spirit of overcommunicating following a boom, Matt can give short but informative SITREPS to his boss and others.

Along with adopting formal formats used for communicating under duress, it is important to designate a person or body (team, department, PR agency, etc.) to manage communication, at least initially as it unfolds. Matt, in the scenario above, could be that person — meaning that Matt can be designated as the point person for information as reports flow in and things change. Matt's boss could announce this to the team to keep reports and information centralized. This does not excuse others from reporting directly to leadership, when needed, but it helps keep information in a central place, such as a shared file on a company server, for all to access.

Communication is everything so let's keep practicing and learning, shall we? Here is another "during-the-crisis" communication example to help you.

A computer virus is detected on a marketing director's computer. Nora is informed of this and, while acting calmly, directs immediate action to disconnect the director's computer and notify Mike in the IT department.

> "Mike [IT leader], give me SITREP please."
>
> Mike uses a 5Ws-plus-next-steps format.
>
> "Boss, Sarah, my IT assistant [*who*], is quarantining [*what*] all seven hard drives of the marketing team [*who*] at this moment [*when*] in the company tech room [*where*] to prevent the virus from spreading further [*why*]. Her expected completion is by 6:00 p.m. this evening [*when*]. She will then transition to replacing those machines from hardware inventory, and I'll report when this happens along with any problems she discovers. I expect we will be up and running again by opening tomorrow. James has formed a CMT to assess the greater impact of this. I'm a member of his CMT."

There are certainly more questions and actions to be taken, such as analyzing the impact to marketing and cash flow, determining who else may need to know, etc., but this is an excellent start.

Let me offer one more scenario. A company car has crashed. Ashley received the call and has been at the center of this crisis from the start.

> "Ashley, please give me a SITREP about the car crash."
>
> Ashley incorporates the 5 Ws in her response.
>
> "Boss, all three of our teammates [*who*] are in ICU but are stable following the car crash [*what*]. Families [*who*] have been notified [*what*] and are either there at the hospital [*where*] or on their way [*when*]. We have a company liaison there to assist families and keep us informed [*why*]. The company vehicle [*what*] has been recovered from the scene and is located at Miller's Autobody, where it is locked up awaiting final investigation."

As we close out our discussion of what to do when things go "Boom!" — and especially how to communicate during crisis — let me share a word or two on transparency. Transparency is vital during the boom and the best way to ensure you achieve that level of transparency when all hell breaks loose is to be planning for it during moments of calm. A spirit and culture of transparency is formed and fostered as leaders build their crisis-ready foundation. Transparency often happens organically as people react to what they see, hear, and experience. However, transparency is not automatic and it's not always natural. In fact, individuals and teams may fall silent in shock, embarrassment, or fear after a boom. Indeed, leaders must form and foster a culture of transparency underscored by psychological safety both in building a crisis-ready foundation and in the left-of-the-boom space. When crisis hits, leaders should ask for transparency as the organization and its people react and as events unfold. Transparency must come from the top and the bottom. Leaders, acting vulnerable, should confess their information gaps and even doubts and subordinates should do the same and report what they discover, even if it is further bad news or exposes them to criticism for their own actions or failings. The only way to be transparent as a company is to be transparent everywhere — up, down, and across the organization.

Battlefield Tip: Have Your Crisis "Cheat Sheets" Ready and on Display

"Share information to the point it makes you uncomfortable." General Stanley McChrystal shared this philosophy in his book, *Team of Teams*.

I've always operated as a transparent leader. I know it helped, especially in crisis. McChrystal codified this behavior for me long after my departure from the battlefield in his book. He is not prescribing an extreme, total level of transparency. Indeed, he advocates some level of discipline to manage, preserve, even hold information as necessary, but lean toward sharing it for the benefit of all. In the end, you (the leader) and your greater team will benefit.

Carry this battlefield tip in your kit bag and extract it when the boom hits — no matter how ugly it is or how liable a person (even you) may be. Get uncomfortable!

REFLECTION QUESTION

Corporate secrets. Are you hardwired for transparency or for information-hording?

The best time to think about transparency during crisis is long before the crisis strikes. But sometimes you must enact radical transparency in the midst of the "boom." Take some time to ask yourself and your colleagues: What could we be doing better — right now — to ensure that we're all comfortable and practiced being open and honest during a crisis? How do we ensure that everyone from our executives to our entry-level works know that they are safe to speak up and share during exigent circumstances?

Is It Over Yet?

It can be difficult to determine when the boom is over. It is not a clean phase with a start and ending. My best framework for determining when the boom is over — when to move on to right-of-the-boom activities — is the term and effect of the crisis having "stabilized." When the situation is stabilized — when lives are no longer at risk, when leaders have control over the situation, when some semblance of order is established, and a solution is underway — an organization can call the boom complete. I don't believe it is proper for a leader to announce, "Boom complete!" Just like an earthquake, a crisis can have aftershocks that are unexpected and that linger longer than we'd hope. Instead of declaring a crisis to be over, leaders should keep watch over the trauma and emotions of their people. While there is much learning to be done after a crisis, it is not proper to huddle the team to review what happened to capture lessons when emotions are still running high. Leaders should focus on consoling their people and assuring them that everything will be OK before shifting to post-boom activities. Leaders should assemble in a small group to analyze the situation moving forward in an informed, reasoned, and principled manner.

CHAPTER 6

Right of the Boom

After the Crisis Has Ended

A crisis, especially a substantial one, represents a crucible — a serious and sometimes severe and difficult test — for a person and an organization. And leading through those crises is inordinately challenging.

Robert J. Thomas in his book, *Crucibles of Leadership*, tells us:

> "In a leadership context, we can think of a crucible as a transformational experience from which a person (or an organization) extracts his or her 'gold': a new or an altered sense of identity ... crucibles are more like trials or tests that corner individuals (or organizations) and force them to answer questions about who they are and what is really important to them."[25]

This alternative approach to thinking about, reacting to, and, indeed, leveraging crucibles found within crises underscores a leader's and an organization's quest to learn and grow from them. Crisis is rarely enjoyable when it is underway, but it can always offer meaningful learning and even change us in profound and positive ways.

---

25 Robert J. Thomas, *Crucibles of Leadership. How to Learn from Experience to Become a Great Leader*, Harvard Business School Publishing, 2008.

A Word About Trauma

While right-of-the-boom activities (i.e., what happens after the heat of the crisis is over) are characterized by learning and growing, leaders must not overlook the effects of crisis on people and organizations. One thing we all have in common is our desire to bury the ugliest chapters of our organizations and our personal lives. Nobody wants to revisit these stressful and horrific moments. Unfortunately, this is impossible ... because crisis changes us. I've done some study on post-traumatic and military-transition stress, having experienced some of it myself. Indeed, there are extreme instances of both — a gruesome combat experience where people were badly maimed or killed, where bad decisions may have led to the tragedy and may haunt a person in their daily life thereafter, or a servicemember who identified so greatly with his or her rank and branch of service who suffers a sense of disconnection once transitioned. And, unfortunately, business leaders are not immune — trauma affects more than just military servicemembers. According to a study published by the National Library of Medicine, more than half of all adults — 61% of men and 51% of women — report at least one traumatic event in their lifetimes.[26] If you lead — on a battlefield, in a boardroom, on a factory floor, in a cubicle farm or an actual farm, in an operating room or a retail storefront, or anywhere else — you and your organization might be the "where" for the traumatic "what" that changes someone forever. It's a heavy burden to bear and a big idea to imagine.

One thing we all have in common is our desire to bury the ugliest chapters of our organizations and our personal lives. Nobody wants to revisit these stressful and horrific moments. Unfortunately, this is impossible ... because crisis changes us.

26 Kessler, R.C., Sonnega, A., Bromet, E., Hughes, M., Nelson, C.B., (1995), "Post-traumatic Stress Disorder in the National Comorbidity Survey, *Archives of General Psychiatry*, 52, 1048–1060.

In my studies I came across a wonderful book by Bruce D. Perry, MD, PhD, and Oprah Winfrey titled *What Happened to You? Conversations on Trauma, Resilience, and Healing*. "Happened" is the operative word, as the book's premise is that what happened to you — especially at a young age when your brain was still developing — effects how you operate in work and life today ... and that your response (ducking for cover at the sound of a car backfiring, an increase or de-regulation of your senses, and the like) is completely natural. Far too many people fail to deal with post-trauma effectively and/or understand why they respond to sights, sounds, and smells as they do, and it manifests in stress, depression, and personal struggles, such as domestic abuse or substance addiction. A business that fails to understand and address the after-effects of a traumatic event — even a simple, yet costly mistake — can suffer the same fate. Organizations are made up of people, and if people can be changed by PTSD, so can your company.

Once we are "right of the boom," we need to invest in understanding and addressing what has happened. This means recognizing the trauma, caring for the wounded — physically and mentally — and allowing time for grief and recovery. It also means revisiting the crisis to learn from it. The mere mention of a traumatic event that had a negative impact on a business and its people may come off as insensitive; but to not do so means the greater organization (and eventually its people) suffers. Under my people-centric leadership approach in my small businesses, I tell my employees — my people — at the outset that, while I care deeply about them and their needs, the business doesn't. My point is not that I, like my customers, don't care. My point is that I and we, as a greater organization, have an obligation to learn and grow and that this will only happen by revisiting the boom. For instance, customers don't care that Joseph is in the middle of a bitter divorce. They don't care that Beth was up all night nursing her sick dog — nor should they. They want outstanding service and quality regardless of personal issues. It is my duty as a leader — indeed, our duty as a business — to strike a balance between caring for and catering to our people and serving our customers. Moreover, the business must strike a balance between care for its people and the hard, necessary work required to learn and grow from crisis.

Leaders and organizations who attempt to bury organizational trauma are doomed to repeat it. They may become timid and risk averse (which, of course, creates different kinds of risks). A sales team may avoid the risk of a product or service pitch following a failed attempt, settling for stagnation rather than going for growth. A business may bring itself to a crawl while examining every part of an activity to prevent a repeat failure as its competitors continue to outpace them; too scared to move quickly, they get left in the dust. Again, people and organizations naturally want to move to a better future rather than revisit an ugly past (that they can learn from) and they will, unless led properly. This is where leadership comes in. I've said it before, and I'll say it again: it's lonely at the top. Many leaders exist in an echo chamber where they, alone, wrestle with their doubts and failings and those of their team. Leaders, even acting alone, must be the ones who muster the courage to bring pause to the organization to 1. Define failures, 2. Identify causes, and 3. Foster learning, growth, and resilience. I'll cover these critical crisis-recovery-and-learning steps in the paragraphs ahead by offering some essential tools for the right-of-the-boom space.

Leaders and organizations who attempt to bury organizational trauma are doomed to repeat it.

The After-Action Review (AAR)

After-Action Reviews or AARs — whether formal or informal — are a requirement for any organization, no matter how large or small, if that organization wants to learn, grow, and survive the next crisis. They are the decisive event in the right-of-the-boom space. Without AARs and the corrective action(s) that should follow them, all is lost. AARs can be elaborate, choreographed events conducted by an external professional like a crisis consultant or they can be impromptu huddles guided by a single leader. No matter their form, they must be done, and it takes courage, discipline, and leadership. Because AARs are so important, I'll spend some quality time here explaining them and preparing you to conduct them.

Let me first say that After-Action Reviews require a spirit of cooperation and learning. Leaders must set a non-judgmental, non-critical tone for AARs to succeed. Additionally, AARs require a Crisis-Ready Foundation, as we discussed in Chapter 3. I'll start with a comprehensive description of the AAR, then I'll simplify it for your use. As with everything in this book, you should consume this part in full. You'll be able to simplify it in execution once you understand it in theory.

After-Action Reviews require a spirit of cooperation and learning. Leaders must set a non-judgmental, non-critical tone for AARs to succeed.

The U.S. Army uses After-Action Reviews or AARs extensively in both training and combat. *Army Doctrine Reference Publication (ADRP) 7-0: Training Units and Developing Leaders,* states, "After action reviews provide opportunities for units to develop critical thinking in leaders." *7-0* defines the AAR as "a guided analysis of an organization's performance, conducted at appropriate times during and at the conclusion of a training event or operation with the objective of improving future performance. It includes a facilitator, event participants, and other observers." *ADRP 7-0* goes on to describe how AARs identify strengths to be sustained and weaknesses that need to improve and that AARs are best conducted throughout an operation at appropriate times to allow for immediate, in-stride corrective actions. AARs are not critiques, the publication stresses. They are part of an open learning environment where facilitators, participants, and observers freely discuss successes and honest mistakes. AARs help units and individuals understand what went right and why, what went wrong and why, and what to do better in the future. I've seen AARs done formally with professional observers, sharing statistics and videos showing the activity of a unit and its leaders. I've also seen AARs done very informally by a leader and a group of subordinates discussing what went right and why and what went wrong and why and how to fix it.

AARs can and will be contentious without a proper organizational foundation and/or if a leader fails to establish the rules. Egos will emerge, blame may be shifted, and people may become defensive. Comments such as "That wasn't my fault," "They [person or group] caused that," or "I have no clue how that happened" are counterproductive. If an AAR drifts into a finger-pointing "blame game" meeting, all is lost. All of us have failures. All of us should have "known better." All of us make mistakes. The AAR is not about assigning blame; it's about fixing problems. Use this chapter to form your guidance for the AAR, setting tone for your organization.

Be Not Afraid

While AARs are naturally focused on what went wrong, there is and should be space to review what went right (and *why* ... and how to sustain that). Time may be short, in which case "what went wrong" should be the focus; however, in the spirit of capturing and sustaining the good, leaders ought to invest time examining and documenting what went right. It is entirely possible when revisiting the ugliness of a crisis event that "right" gets lost. Look for it, see it, and acknowledge it. That said, you're analyzing a crisis, so "what went wrong" surely outweighs "what went right," so I will caution you here: Do not let "right" dominate the AAR. If it does, that is a warning sign that the team is too cautious or afraid to address what went wrong. Sometimes, focusing too much on what went well is sugar-coating the crisis itself. Strike a proper balance.

Come back with me and my soldiers to Afghanistan for a moment. We were dispersed among several bases, operating at the speed of light, when Devoe was killed. It was hard to assemble a group all in one place at one time, so we assigned an investigating officer to interview soldiers and leaders and create a formal report documenting this tragic loss of life, what contributed to it, and what could be learned from it. The report came in the form of "findings and recommendations." Again, it was not an "assign blame and punish" report. We found no malice or deliberate negligence. It was a hardened combat formation against a determined and committed enemy, and the enemy got the best of us — as they were sure to do again.

Even in business — especially in multinational businesses and in the 2020s, with more people dispersed among home offices and remote locations — it can be hard to pull everyone together for a face-to-face, synchronous meeting for an After-Action Review. Whether or not you can gather your team, be sure to conduct some form of an AAR, formal or informal, in person or virtual, every time there's been a crisis. Here is an agenda and some ground rules, which you can use for the conduct of your AAR.

AFTER-ACTION REVIEW (AAR) AGENDA

- **Scope.** Identify the major events to unpack for the AAR. It is helpful to focus the AAR on specific moments and events to prevent too broad a conversation. As a leader, spend some time thinking about the areas (events, places, times) where you believe mistakes occurred, and where you think the organization or individuals could do better.
- **Objective(s).** What did we set out to do? What was the mission, intent, or goal at the start? What guidance was issued for it? You may need to return to a founding document or policy that guides the organization or activity. If there was no guidance or directive given, force something from the group. What did you/we imagine the outcome to be when we started?
- **What happened?** Avoid sprinting toward why it happened or what could have been done differently. Focus instead on telling the story of what happened with all its faults. Just get the truth out there before you begin to dissect it. Again, avoid placing blame or singling out anyone.
- **Why did it happen?** This is where root causes and lessons begin to appear. Again, avoid rushing to what can be done in the future before you and everyone on the team hear and understand the full story from start to finish. Here you can make notes of things you believe the team should address.
- **What can be done to sustain the right and fix the wrong?** Only after the above agenda will this portion of the AAR work. Here is where you and your team harden yourselves for the future. Remember this — assign responsibility, or as I like to say, pin the rose. The lessons teased out of an AAR may go unlearned unless the proverbial rose is pinned on a person or entity to take concrete steps to prevent them from happening

again. A department or section leader or director (or even an individual with specific expertise, such as technology) can receive the rose.

As you are able, document the AAR for future use, even if all you do is record it on a phone or snap a picture of a notes taken. Have someone, you even, capture the lessons learned for all to reference. This can be done in a page or two of the main points and remedies the team will abide by in the future.

HOW TO PREPARE FOR THE AAR

- ❑ Find the right space away from distractions. Remove cell phones and require the full attention of members.
- ❑ Designate a "scribe" or note taker and, if needed, a timer so that key points are captured and time does not drag on.
- ❑ Create and be prepared to issue the rules of the AAR. Be the arbiter of them. Here are rules you can use as a starting point:
    - Everyone has an equal voice.
    - No comments are to be ridiculed.
    - Have thick skin. Don't take criticism personally.
    - Focus on learning, not defending oneself or blaming others.
    - Leave excuses outside the room. Open up. Confess where you may have gone wrong.

After the AAR

Conducting the AAR does not guarantee no repeat mistakes. The real insurance against repeat mistakes and future crisis is what we do *following* AARs. AARs can be fantastic events which, as the U.S. Army described, develop critical thinking in an open learning environment. There is goodness in conducting them; however, their efficacy is diminished without some form of corrective action following their completion. While the Army was very disciplined at conducting AARs and documenting their lessons, it was less effective at implementing changes. It did, indeed, act upon the lessons of the IED but, on a smaller scale, following AARs, many Army units would undergo a significant change in personnel and leadership or return to a chaotic, overloaded schedule. After some AARs, I was guilty as an Army commander for lack of follow-up. In the AAR, where all of us were still reeling from the recent crisis — bloodied and battered, where I had the attention and commitment of all participants — we would capture lessons and I would "pin the rose" on individuals to take on the "fixes" we teased out of the review. In some cases, I did this knowing that the fix(es) may never take hold. The lesson here is to determine as a leader, with help from the audience, which roses need to be pinned and upon whom — which person or team will take the right-of-the-boom learning forward, where it can improve our operations in the future.

There is goodness in conducting them; however, their efficacy is diminished without some form of corrective action following their completion.

Leaders can (and should) assign "homework" for AAR participants to develop policies, standard operating procedures (SOPs), or formal or informal rules for the organization and its current and future members to follow. Don't fear the word policy or the acronym SOP. They get a bad rap because they are akin to the thick rule books or laws we must follow in our personal and professional lives, especially ones created by our government (like the tax code!). Most of us must hire professionals just to understand and abide by them. Policies and SOPs can and should be short, simple documents — a few words even — which people can read and understand under the madness of overloaded schedules and "to-do" lists. There is a simpler, yet effective, way to do this.

Let's look at that IED blast in 2009 through the lens of the AAR agenda. The **scope** for Devoe was broken into phases — preparation, the operation, and actions following the IED. We wanted to know what we did right and wrong in each of these phases to better ourselves. Our **objective** was three-fold: 1) Route reconnaissance. We needed to familiarize ourselves with the terrain and a primary route and area we would operate in. 2) Kill or capture enemy. 3) Engage with the population. Under **"What happened?"** leaders described in a story format what they did during each of the phases. In the preparation phase, leaders did their typical pre-combat checks and inspections. This included assignments of roles, such as drivers, gunners, and patrol leaders. Radios were checked and weapons test fired. Intelligence was studied and shared, and routes examined. It was here where we found that we could have done better work planning our route and studying prior intelligence to determine where the hot spots (areas where attacks previously occurred or IEDs were detonated) — our **Why did it happen?** Under the **What can be done to sustain the right and fix the wrong?**, we could have dismounted our vehicles at different spots along the route to detect and avoid IEDs.

The lessons from Devoe's death were implemented across my task force, especially along the route that Devoe's patrol traveled. Among other lessons learned, we dismounted more frequently when traveling in high-risk areas. Heeding the lessons of the previous unit, we established a remote patrol base — named "Devoe" — to strip away the enemy's ability to fire upon us where they had before. It worked. There were no more casualties in that area during the remainder of our tour.

Now, let's take this discussion from the desert to the dining room — from examples in military combat to the far-less-stressful world of interior window fashions (one of the businesses I own and operate in my post-Army civilian life). In my business, All About Blinds, for our **scope**, we selected events where we were sure an error occurred, such as the in-home consultation, the measurements of the windows, or the steps taken to enter the order in our system. Our overall **objective** was to understand customer desires, conduct an accurate measure (i.e., get all the dimensions correct when measuring windows for new blinds, shades, drapery, or shutters), and enter the window-dressings order without fault — to ensure a seamless and flawless transaction. Under this objective, our goal in the AAR was to tease out the flaw(s) and address it in detail. Our "**what happened**" turned out to be human errors. There were incorrect measures, where we didn't account for factory deductions in width. We sometimes forgot to account for obstacles preventing smooth operation or we ordered a product requiring a different measure. Sometimes the error occurred between customer expectation and what we delivered because the customer was not engaged or present during the consult/measure. In terms of **why these happened**, we discovered that we were moving too fast, lacked proper product knowledge, and/or failed to revisit the customer to ensure they were getting what they actually desired. Under **"What can be done to sustain the right and fix the wrong?"**, we first developed a single-page document with all the most common risks that caused failure. This was a bulleted list of potential failures at the point of consult where a customer expressed doubt, was preoccupied with something, or was not clear on how the product would operate or what it would look like. When we encountered one of these risks, we would plan a second visit or a detailed review with the customer. I altered our ordering process, involving more of the team to review the order before submitting it. Lastly, we spent more time with our distributor, understanding the product and its specifications. "Leave no doubt," became our mantra; a sentence which captured the essence of what we needed to do — a sentence born of the AAR in the right-of-the-boom space.

Keep It Simple, Crisis Leader

Simple, yet impactful policies, SOPs, words, or statements are the real saviors in the right-of-the-boom space. I'm reminded of the phrase "Measure twice, cut once." Mnemonic devices foster information retention or retrieval (remembering) in the human memory for better understanding. I'm guessing you use mnemonics all the time, like the "ABCs" song to remember the 26 letters of the English alphabet or abbreviations like "Roy G. Biv" to remember the sequence of the colors of a rainbow (red, orange, yellow, green, blue, indigo, violet). We forget things as human beings, and this becomes more prevalent and more dangerous when a business is seemingly moving at the speed of light. Organizational values are a form of mnemonics — single words that have deep history and meaning, and that guide a business and its people to successful and exceptional outcomes. I wrote "Leave No Doubt" on our company dry erase board where we would write frequent notes reminding us of things to be done. We could not avoid seeing it as we returned to the board to revisit our tasks and write new reminders.

In the right-of-the-boom space, it's important to get our thinking organized and our information simplified into manageable formats and quantities. We can do that thinking as information "chunking" — a form of mnemonics. It's about breaking down information — sometimes overwhelming and complex information — into bite-sized "chunks." Two common examples of chunking are phone numbers and Social Security numbers (SSNs). Most people divide these long numbers into three sections (123-456-7890 for a US phone number and 987-65-4321 for an SSN), in essence learning them in chunks before placing them together. The "Left and Right of the Boom" concept itself is an example of chunking. It might be difficult to learn and retain all the ideas in this book of I presented them to you in a giant list with dozens of frameworks and tools and no organizing principles. But when you think about crisis in three phases (left, boom, and right), you've begun to simplify when and how to improve your prevention, leadership, and response to crisis. Crisis response and planning can be further divided into specific activities done inside each phase of managing crisis. "Before, during, and after" (the words I use in this book's subtitle) is another way to use chunking. "Given what we have learned, what are the top three objectives we must accomplish before, during, and after our next mission?"

REFLECTION QUESTION

What are your favorite ways to organize information into chunks, buckets, columns, categories, etc.?

If you work in healthcare, you might think about organizing your patients into "new patients," "wellness/preventive patients," "acute patients," and "chronic patients." Or if you run a bookstore, you might think of your inventory as being "classics," "new releases," "back-list titles," "book-club favorites," and "local authors." Take a minute to think about the many ways you "chunk" up or organize information at your company and keep it simple.

Start with titles (buckets) like the above examples, then think about what goes in them. Then ask yourself, "What's working in that regard and what's not?" Where do you need to define and communicate these proverbial buckets better? What chunks, buckets, columns, categories, or other bite-sized containers can you and your colleagues start talking about to create a common vocabulary and some organizing principles to ensure your work is more effective and efficient (and that nothing slips through the cracks)?

Acronyms are another great example of mnemonics. The Army was famous for acronyms, but it had too many. Be selective. We needed mnemonics just to memorize all the Army acronyms! Here is a humorous (yet effective) one for your kit bag. The values of my small business are Selflessness, Humility, Innovation, and Teamwork. If you didn't immediately pick up on that, it spells SHIT. That acronym was not my intent when creating the company values, but it's funny and undeniably memorable. And maybe focusing on SHIT keeps us from getting caught in too many shitstorms.

Rhyming and music are other effective forms of mnemonics; ask any professional marketer about the power of a jingle or slogan, and they'll start singing and telling dozens of success stories. Who can forget "Yo quiero Taco Bell!" or "Nationwide is on your side?" The commercials touting these slogans — often delivered with music — are designed to make a permanent imprint on our brains to guide our shopping activities. So too do the phrases "thirty days hath September ...," "rightsy tightsy, lefty loosey," and "measure twice, cut once." These simple, yet memorable, mnemonics can save the day when a group or individual is in the bustle of the moment yet must prevent failure. I never knew it in a formal sense, but I used mnemonics in leadership. I still do, especially after an AAR. An old boss of mine once said, "If you want something done right once, you have to say it 10 times." I find, especially in the right-of-the-boom space, that the more repetitive I am — especially with short statements or words — the greater chance the true lesson or lessons from the AAR sink in. "Leave no doubt" became our "Nationwide is on your side." I even started hearing it back from my people, an indication that it was sinking it. It certainly did not mean "problem solved," but it was and still is an essential activity in my right-of-the-boom space.

I find, especially in the right-of-the-boom space, that the more repetitive I am — especially with short statements or words — the greater chance the true lesson or lessons from the AAR sink in.

As you lead your team in the right-of-the-boom space, a nice way to conclude it or transition from it is with a SWOT analysis. SWOT is an acronym for Strengths Weaknesses, Opportunities and Threats, and a SWOT analysis is an assessment every business must conduct and revisit from time to time if they want to remain competitive and resilient. On the right side of the boom (and, more specifically, following the AAR), a SWOT analysis is an effective way to refocus the team. We did not use SWOT formally in the military, but we certainly reviewed its elements as we reflected on ourselves, our teams, and our missions.

No matter what kind of work you do and in what type of industry, a business can evaluate itself across functions and attributes — such as product or service, marketing, finance, management, and operations — using a SWOT analysis. And if you're new to the concept, you can download a template online.[27] SWOT requires a heightened level of honesty to look at your organization as it *is*, not as you *wish* or *believe* it to be. While it is important to conduct a full SWOT across business functions and attributes, here I'll give you an example of a generic SWOT for my small business, which capped our After-Action Review (AAR) and helped us see ourselves and ready ourselves for the future. Remember to put a time stamp on your SWOT analysis. Under changing environments — where opportunities appear and threats materialize — the shelf life of today's SWOT will soon expire, and you don't want to be guided by a stale and irrelevant SWOT analysis.

Remember to put a time stamp on your SWOT analysis. Under changing environments — where opportunities appear and threats materialize — the shelf life of today's SWOT will soon expire, and you don't want to be guided by a stale and irrelevant SWOT analysis.

27 SCORE Association, "SWOT Analysis Worksheet," April 10, 2024, https://www.score.org/resource/template/swot-analysis-worksheet.

2024 SWOT Analysis: All About Blinds	
STRENGTHS My business carries a time-tested window-fashions brand called Hunter Douglas — a 100-year old company with a rich history as an industry leader. I could ride that reputation every day. Amidst our faults, we had done yeoman's work to earn some great Google reviews and we stood upon the great reputation that the previous owners of "All About Blinds" had built. Anyone assessing my business on the internet would not see the internal struggles we were experiencing — they would only see the rave reviews our customers left. We are well led and have talented and dedicated team members who always rise to the occasion and carry the day. Finally, we are in a growing region in terms of new home construction and real estate performance. These homes need window fashions, and we're here to help.	**WEAKNESSES** We were clearly lacking experience and acumen in the sale and installation of window fashions. Most of us, when we joined All About Blinds, were new to the industry. While I had options to raise money, we didn't have the cash reserves to weather the storm of failed installations, a pandemic, or an economic downturn. Recently, our expenses were too high. I had made some advertising commitments that were costly and not producing results. Finally, while we were a small team, we were a bit segregated in our duties. We were failing to come together at key moments to share information, help each other, and ensure a near-flawless operation.
OPPORTUNITIES The opportunity to formally address our failures was the central golden opportunity facing us. We had nowhere to travel but upward to a greater place. I revisited "what can be done," from our AAR to reinforce our commitment to bettering ourselves. There were indeed opportunities external to our business. These opportunities could come in the form of relationships and partnerships with designers, realtors, builders, and other professionals in the home building, selling, and improving space. For our purposes in the right-of-the-boom space, I focused internally on the opportunities to better ourselves. These opportunities would manifest in simple practices, such as an extra check on our measurements and product specifications before pressing "send" on an order and videos we could watch from our manufacturer to sharpen our methods. We documented lessons (opportunities) from our failures such as "window measurings gone bad" as a mnemonic to remind us and to better us.	**THREATS** Of course, the immediate threat was the blowback - bad customer reviews, refusal to pay, etc. — of our poor performance. Immediate steps were taken to mitigate risk of repeat failure. However, complacency — a self-imposed, internal threat — would remain our greatest threat. The energy and motivation were high to address our faults and better ourselves following recent failures, but I knew over time we would revert to our old ways. I did visit external threats, such as competition and the economy but to keep focus, I stayed with the current crisis, choosing to control what I can control right now.

Moving On

As I said before, leaders and organizations naturally desire to package up failure, place it behind them, and move forward. I'm no different and neither are the organizations I've been a part of are, but it is incumbent upon a leader to peel off the bandage and revisit old wounds to foster learning and growth. I'm talking about fostering an enduring environment of learning and growth. This is done by revisiting past errors and their lessons. It can and should be one at key moments where the team is gathered or when a person or group is about to engage in an activity where previous errors occurred. That said, it can be harmful to morale if your organization has someone who is always (and only) focused on what went wrong (and not what went right), so leaders must be judicious about when they revisit the ugliness and the pain. If it is done too often, trust may begin to erode. My radar is constantly scanning for times and events where I think an error might occur. My people know I trust them, but they also know I'm there to help them stay sharp, focused, and error-free. That scrutiny goes both ways; I'll ask them to check me when I do something, as I am human (and error-prone) too.

It can be harmful to morale if your organization has someone who is always (and only) focused on what went wrong (and not what went right), so leaders must be judicious about when they revisit the ugliness and the pain.

Following our efforts in the right-of-the-boom space in my small business, we were able to overcome our flaws. It was wonderful to watch but I always kept my guard up. One day, quite spontaneously, I stopped the team. "Ok, what are we missing? Things are going too well. What are the risks and what are the faults we are not seeing? What could fail in this moment and what can we do about it?" While my words silenced the team, I got them thinking and revisiting old errors. We teased out a few things and it refocused us into being as sharp as we were when we were fresh from our previous mistakes.

One Step at a Time

As with other phases and periods of crisis leadership and management, you should treat right-of-the-boom activities as separate and distinct. Announce to your team that you are all in the right-of-the-boom space and describe what activities will occur there. Discipline yourself and your teammates to participate in reflecting and learning for the betterment of your individual and team performance and the whole organization. Reinforce your confidence in your people and explain yourself often, so that you are not misunderstood by asking hard questions. Be personally vulnerable. It is proper to admit fault and accept responsibility. By doing so — by showing you are human — you inspire your people to follow and emulate your behavior. Remember that average organizations, even bad ones, bury their mistakes and run from them. Great organizations invest time understanding failures, learning from them, and growing — *right* after things go Boom!

CHAPTER 7

The Tactical Operations Center

A common theme with any organization in crisis is the need for information, command, and control. Risks are high, information travels at the speed of light, and key decisions — guided by the best information possible — need to be made. Here, I introduce the concept of the tactical operations center or TOC. A TOC might seem unrealistic to a small business, which (especially under duress) cannot afford the people, time, or resources to form and run one. Stay with me. There are loads of value in this chapter. I've included this chapter because, almost daily, I rely on my experience working in and running a TOC to guide me in crisis. The doctrine to follow helps me receive, process, and distribute information and lead more effectively, crisis or not. I'll start by defining a TOC then segue to its principles to help you in all phases of the boom.

In the military, commanders turn to assembled bodies of people and systems or TOCs to help them to "see" the battlefield (crisis), to react, and to guide their people and organizations appropriately. The TOC is the central hub of an organization where leaders can conduct informed command and control, gathering and processing information, providing or reallocating resources, directing subordinate people and teams, and managing crisis. Military units establish and staff TOCs or command centers in training and in combat to run operations 24/7. Combat never sleeps. Crisis doesn't either. There are simple, yet useful, ways a small business can replicate what a TOC does for its own crisis management. Even the slightest imitation of the concepts outlined in this chapter may save the day when the chips are down.

Military units establish and staff TOCs or command centers in training and in combat to run operations 24/7. Combat never sleeps. Crisis doesn't either.

In crisis, the need for command and control is essential. "Command," in military parlance, refers to the authority given to commanders (or CEOs) and how they use this authority for the betterment of their people and their organizations. "Control" is the art of guiding people and teams through specific directives toward a desired end state. *U.S. Army Field Manual (FM) 3-0: Operations* offers a concept called "Battle Command — Understand, Visualize, Describe, Direct" as an effective way to command and control. The role of commanders or leaders, especially in crisis, is to exercise battle command as it applies to their people and organization. Good leaders do not lead erratically without some analysis and forethought. They spend time, even if only seconds or minutes, understanding and visualizing what is happening in the moment, how it is related to the mission and an envisioned future, and what risks there may be. They ought to verbalize it, even if only to themselves, to better direct their team. Once leaders have visualized, they then can describe the situation and its problems. Once the key steps of "visualize" and "describe" are complete, leaders can "direct" with greater effect.

My visualization of Devoe's death was that we were a new unit being tested by the enemy. We were under attack and needed to take immediate action to protect the remainder of the force. I described, largely for myself, that my unit was tense, eager, and in a state of shock while operating for the first time, after a year of training in this faraway land, a combat zone. We were hungry to take the fight to the enemy, to learn everything we could about the terrain and the population, and to become experts at warfare. I further described that, while most in the formation were new to war, I had seasoned leaders at several echelons who knew what to do under fire. I sensed the emotions of my team and how — without calm, confident, and trusting leadership — they could let this situation impact, in a negative way, their behavior. My direction was unemotional and simple. Subordinate leaders would execute the proper tactical steps — securing the area, recovering Devoe, and returning to base. I directed commanders to tend immediately to those in trauma over the loss of their fellow soldier and to keep their unit focused on the mission. We were facing a full year of danger.

In my small business, where things are less volatile and where crisis is rarely a life-and-death proposition, we still experience setbacks. And when we do, I still fall back on the "understand, visualize, describe, and direct" approach. After a series of setbacks at my company in its inaugural year, my visualization reminded me that we were, after all, in "year one" of the business and relatively new to this industry. I described my team as eager to sell and install window fashions and move on to the next client. I believed, because we had not yet experienced failure, that we lacked the attention to detail demanded of our profession. We may have been a little cocky. My direction consisted of halting the ordering of all products without my approval and requiring team reviews of orders.

Back to "command and control." A common theme across the elements of Battle Command or commanding and controlling is the need for information. In most Army operations, commanders rely on tactical operation centers (TOCs) to help them see and think clearly. Keeping in mind that a TOC in a small business might be less of a "center" than a single person with the right mindset (and that person might be an operations or office manager, or even you, the leader), let's dig deep into what a TOC does and how you can replicate these concepts in your own crisis-management endeavors.

Army Techniques Publication (ATP) 6-0.5: Command Posts and Organization and Operations governs the establishment and operation of a TOC. A simpler, more digestible, description of a TOC and its functions can be found at GlobalSecurity.org. "A tactical operations center (TOC) is a command post for police, paramilitary, or military operations.

A TOC usually includes a small group of specially trained officers or military personnel who guide members of an active tactical element during a mission." Here are the essential functions of a TOC:

1. Receive information
2. Distribute information
3. Analyze information
4. Submit recommendations to the commander
5. Integrate resources
6. Synchronize resources.

1. Receive Information

Again, the TOC serves as the central hub of information. It establishes direct and often continuously open lines to other key individuals and teams to gather information, helping the commander or leader "see" more clearly. The TOC usually has a primary way (and several alternate ways) to contact and communicate with an individual or team — it is that important. Coupled with methods to communicate, the TOC maintains the location and activity of individuals and teams so that it knows when and when *not* to communicate with them. Receiving information involves much more than having a person's cell phone number.

TOCs also determine what *type* of information is required. They establish a regimen of reporting to maintain real-time situational awareness and they update the information, as needed, when the operational environment changes. There may be a report like a daily update, which includes location, activity, and assessments (SALUTE style). There will likely be information requirements, such as engagement with the enemy or when a critical class of supply (like fuel or water) reaches critically low levels. In a business, reporting on things like the volume of "cold calls" to sales prospects, contract conversions, and job closures may be information required for the leader to "see" more clearly.

2. Distribute Information

Information travels on a two-way street. Poorly operated TOCs misunderstand this and fail to keep leaders and teams informed. Recall the methodology of "what is happening and who else needs to know?" in the 5 Cs of crisis leadership. TOCs can and should distribute information that's useful to the greater team. A TOC can typically see things with greater clarity. TOC members generally sit, in comfort, out of the environmental elements in secure, generally undistracted areas where they can think, analyze, and distribute information. TOCs *can* "see the forest for the trees," while others who are in the thick of the forest simply can't — because they're surrounded by the trees. Individuals and teams will not usually enjoy the wider view the TOC has; therefore, it is imperative that the TOC distribute information, letting others know what they can see from their bird's eye view. To use another analogy (this time a military one), imagine how you can see a landscape different from a helicopter than you can from one single position on the ground. And viewpoint, in a crisis, is everything. The TOC is the perfect venue, with all its communication tentacles, to distribute the intent and directives of a leader. A leader, using their TOC to *understand* and *visualize*, can *describe* and *direct* to numerous people and teams so things are optimally handled before, during, and after crisis.

Information travels on a two-way street.

When Devoe was killed, the TOC instantly became the hub of information distribution. Following the operating philosophy of "what is happening and who else needs to know?," TOC leaders informed the appropriate commanders, including me and my commanders. They informed adjacent units, especially those close to the area where the attack occurred, so that they could react appropriately by increasing security and surveillance to identify and defeat threats. A variety of support people and units — such as logistics, medical, and even spiritual — needed to know to prepare evacuation assets for soldiers and equipment. Periodic updates were distributed to keep everyone abreast of the situation. Again, in concert with routine updates and standard reports, TOC members applied curiosity seeking, receiving, and distributing information using verbal and automated means. In the immediate aftermath of the "boom," the TOC was a flurry of activity and communication.

3. Analyze Information

TOCs are not just numb, lifeless information-reception hubs. They can and should analyze information received to help the leader and the greater organization "see" more clearly and act in an informed way. Information analysis is the vitally important "so what" of information. For example, if extra lumber was used on a particular job site for a construction company, their TOC, having received this information, could do the analysis that the next client's job would be short on required lumber. It could also determine that the lumber yard would close before the team could replenish. It could present this information, coupled with analysis, to the leader and the team to facilitate decisions. The best TOCs (and, indeed, the best individuals who run TOCs) have an inherent curiosity to them. They thirst not only for the information they receive but, more importantly, what it *means* and how it connects to and impacts operations. They use this analysis to present recommendations to the leader.

The best TOCs (and, indeed, the best individuals who run TOCs) have an inherent curiosity to them. They thirst not only for the information they receive but, more importantly, what it means and how it connects to and impacts operations.

4. Submit Recommendations to the Commander

I spent a career developing myself and my subordinates to provide recommendations to leaders above us. Most of us are subordinate to someone and the skill of analyzing information and producing well-crafted recommendations to our superiors is a skill worthy of development. A well-functioning TOC does this effectively. It and its people operate with extreme curiosity under questions and philosophies, such as, "Don't bring me problems, bring me solutions," "So what?" and "What is happening and who needs to know?" Using the lumber scenario, a TOC, having gathered the information, could present it in this way: "Boss, team 1 reports an overage in the use of lumber on their second job today. This will impact their third and final job and, at the rate they are progressing, they will not be able to visit the lumber yard today in time to replenish. Team 2 finished early and is one hour from the lumber yard. If we redirect Team 2 to pick up the required lumber for Team 1's final job, they will be able to drop it off in time for Team 1 to complete the job. This is my recommendation." Here is another way — the "bottom line upfront" way to present this. "Boss, I recommend we redirect Team 2 to pick up the required lumber for Team 1's final job. Here is the analysis behind this recommendation. Team 1 reports an overage in the use of lumber on their second job today. This will impact their third and final job and, at the rate they are progressing, they will not be able to visit the lumber yard in time to replenish today. That will cause delays tomorrow. Team 2 finished early and is one hour from the lumber yard so they can jump in to pick up the supplies."

The essence of providing effective recommendations is to see things through the eye of the leader. For example, one of my operating philosophies in my small business is "get the customer to yes." This means we go the extra mile to serve our customers. This approach comes from my military training, serving under senior leaders. It doesn't mean that we push the customer toward saying "yes," such as closing a sale. Rather, we challenge ourselves, within reason, to say yes to the customers' demands, big and small. For a business, this is a competitive advantage. For leadership in crisis, it is a sound application of leadership and followership. Does the customer need a schedule change, needing their measurement, quotation, or installation on Tuesday instead of Monday? "Yes." Can we assist the customer in moving a desk away from the window while we are on site? "Yes." Does the customer have an odd-shaped window that needs a specialty product to fit it? "Yes, we can do that." We will find a way. Indeed, "yes" is not always possible; however, under this operating philosophy, all members will do the hard work to make the customer happy instead of offering a "no" at the first obstacle. Operating under this philosophy, my TOC (my office manager) provides me recommendations in the spirit of getting to "yes." Even if it is not monetarily feasible, they understand I want to hear all possibilities.

5. Integrate Resources

Resources exist in a wide variety of categories, akin to tools in a large toolbox. A good TOC will know the full inventory of resources in the toolbox, use them in providing recommendations, and integrate them appropriately to accomplish missions. Steve F. Krar, an accomplished Canadian manufacturing researcher, author, and educator offers a simple, yet comprehensive, list of resources. I'm sharing them here because I believe they're a useful checklist for you and your organization in crisis. Keep them handy and add to them as needed. Krar's framework organizes resources into these categories: people, information, raw materials, tools and machines, energy, capital, and time.[28]

A resource is anything that adds value or utility toward something or somebody (or a group of "somebodies," like a team). Simply stated, resources are the things we need to get the job done. Using Krar's framework, let's address a few of the critical and most universal resources here.

Types of Institutional Resources

- *People (internal employees and external contractors or consultants)*
- *Information (existing reports, internet searches, AI applications)*
- *Raw Materials (renewable and non-renewables)*
- *Tools and Machines*
- *Energy*
- *Capital*
- *Time (work hours and post-work hours)*

28 Steve Krar, "Technology Resources," *Manufacturing AUTOMATION Magazine*, https://www.automationmag.com/images/stories/LWTech-files/83%20Technology%20Resources.pdf.

People are an obvious resource — a precious one, indeed! It is why they are called human capital (a term you know I dislike if you have read any of my books). I'll add to "people" in the toolbox the concept of the leader. I've always offered myself as a resource to any team I led. Many of my subordinates struggled to see me as a resource and, moreover, felt uncomfortable labeling me or tasking me as such. To help them, I explained it this way: Often it is only the leader who can communicate with another person or entity, or whose influence wields greater influence and power. In my small business, whenever we have a significant challenge with a customer, I can be the resource applied to remedy the situation. Only I have the authority to spend additional funds to please a customer or, if needed, the authority to tell them "no." In a scenario where a customer was on the fence about whether to spend a large sum of money on the products we sell or if the customer was unsatisfied with our product or performance, I could be the "resource" applied to the situation. In Afghanistan, as the TOC worked to integrate resources, they called on me to meet with and influence prominent Afghan leaders. In essence, I was a resource for my organization. Only I — because of my position, authority, and relationships I built — could influence Afghan leaders to the degree needed. I've included a link to a video which offers an example of this in action. Hard as it may be, imagine me as a non-military leader and/or business owner with the same position, authority, and relationships interacting with a stakeholder or customer. In both the combat and small business scenarios, beyond the influence and power I wielded, it sent the right message to have the "boss" as the spokesperson and decision maker if needed.

Come with me to Afghanistan, for some examples of using people — especially leaders — as a resource in getting things done.

Information should be viewed as a *resource* not just as an exhausting flurry of phone calls, texts, email, and conversations that keep us busy each day. A TOC must analyze information to see if it is useful to individuals or teams and to them appropriately integrate it in support of mission accomplishment. When the IED struck my unit and killed Devoe, information became the vital resource to manage the situation. The TOC needed to understand the existing threat so that it could allocate (or not) additional resources to prevent further loss of life and to destroy enemy forces specific to this attack. The TOC diverted attack aviation to provide surveillance and lethal support and it took SALUTE reports from the subordinate command to know what and how much medical evacuation support would be needed. Lastly and importantly, the TOC gathered, analyzed, and provided information for me to understand the situation and make decisions. A TOC or an individual in a business would execute the same steps to manage a crisis, big or small, in a business of any size. It is imperative for everyone to see information as a resource. In support of this, leaders must emphasize and enforce reporting — from the bottom up and from the top down — for their organization to operate efficiently, especially in crisis.

Time is a resource in perpetual short supply. It is this fact, therefore, that makes it such a precious resource. I've lived my life wrestling with time, trying to squeeze as much into a workday as possible and managing a schedule to get the most out of a day, week, or month. In crisis, schedules and events must be analyzed to find opportunities to provide time as a resource to overcome challenges. You've heard the expression "buying time" — in organizations, we need to be looking for ways to do this. Immediate changes can be made to schedules to open time for necessary events to occur. For instance, a company may shift an appointment with a client or pull some people off the job so they can get some needed training done to prevent further problems. When Devoe was killed, we allocated time on our packed schedules to conduct a remembrance ceremony to honor his life and to help a unit in mourning come together to remember, heal, and recommit ourselves to our unit, its people, and our mission. If something is important enough, you'll find the time.

6. Synchronize Resources

Synchronization is complex and difficult. I've met few people who can truly master it — those who can tightly manage it through planning to execution. Again, this is why we have TOCs. The simplest way I know to synchronize is to create a synchronization matrix (or SYNC Matrix), where you stack the actions of people, teams, and resources vertically across time or events to determine where friction points may occur and/or where resources may be unavailable or in short supply. This is done in isolation, away from the friction of the operational environment. From this template, decisions can be made, resources allocated, and schedules adjusted. Once in execution, the SYNC Matrix will serve as a management tool to direct operations in a more informed way. TOCs can more clearly see where elements of the organization are in time and space, so they aren't operating and directing in the blind. Below are two examples of a SYNC Matrix, one for a trucking company and another for an insurance company displaced by an office severely damaged by a fire. This SYNC matrix format is modeled after a military synchronization matrix found in the *U.S. Army's Field Manual 6-0, "Commander and Staff Organization and Operations."*[29] I've simplified it for you.

This SYNC Matrix involves a trucking company's steady-state operations. This product could be the result of a crisis (boom) experienced by the company where they conducted an AAR, discovered they were not synchronized in their efforts as the root cause of the crisis, and used this book to guide them. As you can see below, key periods of time are listed across the top of the matrix. In this case, the trucking company has decided to divide the day into four key periods where they know synchronization is essential. Along the left column are people — responsible agents who own the activities listed in their row. Not all blocks must be filled. For instance, the CEO may be working on future contracts, away from the company during the day. For our purposes here, I've kept this simple. Simplicity and brevity is very important; attempt to achieve that but know that more activities could and should be identified for individuals during the periods listed across the top.

29 *FM 6-0.* https://armypubs.army.mil/epubs/DR_pubs/DR_a/ARN35404-FM_6-0-000-WEB-1.pdf and https://armypubs.army.mil/ProductMaps/PubForm/ContentSearch.aspx?q=6-0

Time/Event	Evening Prior	Morning	Midday	End of Day
Person/Agency				
CEO	• Spot checks vehicle logbooks • Receive Operations Brief	• Greets and inspects drivers		• Present to greet drivers
Dispatcher	• Checks fuel levels • Driver-vehicle assign-ments complete • Secures vehicle lot	• Reviews/ adjusts schedule • Driver-vehicle adjustments	• Route management & adjustments • Vehicle recovery, as needed	• Oversee vehicle return and future readiness
Operations Director	• Finalizes and rehearses delivery schedule • Replacement (rental) vehicles present for next day	• Issues changes & makes adjust-ments as needed • Future planning	• Finalizes future delivery schedule	• Reviews day's events and routes missed
Fueler	• Ensures vehicles are topped off	• Opens and operates fuel point as needed	• Mobile fueling as needed • Ensures fuel point is ready for end of day	• Fueling operations
Lead Mechanic	• Reports inoper-able vehicles • Evening mainte-nance opera-tions as needed	• On site for checks and repairs	• Mobile mainte-nance opera-tions as needed	• Vehicle inspections
Bookkeeper	• Processes fuel and maintenance invoices	• Fuel card checks		• Disburses rental vehicle payments
HR Director	• Produces Driver Readiness Roster	• On site for personnel issues		• On site for personnel issues

For good measure and for good practice, let's explore one more SYNC matrix. This SYNC matrix involves an insurance company whose office caught fire. They have vacated the damaged office to three temporary offices to keep continuity of business. Given the lack of a central meeting place, the company decides to form mobile sales teams in the interim. This matrix uses events instead of periods of time. Indeed, additional matrices could be created to synchronize the re-establishment of the original office. It also is much less detailed than the trucking company matrix, allowing for autonomy and internal work within each function. This company decided to view this disruption as a journey — vacating a damaged office, operating displaced, then transitioning back to the original office. The key milestones of the journey are listed across the top. Instead of people, this matrix depicts functions. This is helpful because it can capture all the functions that make an organization work — logistics, finance, sales, etc. Inside of these functions there are indeed people — leaders with the proverbial rose pinned on them responsible for the activities listed in their columns. Again, this example is simplified. There are, indeed, more activities required across all these functions, and more events to ensure the prevention of future crises.

EVENT	Vacate Damaged Office Occupy Temp Offices	Recovery and Repairs of Damaged Office	Closure of Temp Office Space	Re-establish Original Office
FUNCTION Leadership	• Assist with customer notification and care • Ensure care and operability of employees	• Direct recovery effort • Site visits to temp offices • Receive recovery briefings	• Brief teams prior to move • Develop phased approach to reoccupy office to ensure continuity of customer care	• Establish office evening prior • Welcome party • Assist with customer and employee needs
Administration	• Establish support teams for 3 temp offices • Establish temp contracts for office support (shredding, printers, etc.) • Arrange for phone-line transfers	• Manage support of 3 temp offices • Manage temp contracts • Develop plan for reoccupation of repaired office	• Coordinate phone line transfers • Refresh office cleaning contract • Restock office supplies	• On site to address needs of employees • Relieve temp administrators and terminate temp contracts
Sales	• Customer care surge operations (CRM) • Establish mobile sales teams • Establish private meeting space at temp offices	• Manage mobile sales teams • Measure impact of office move. Adjust sales effort accordingly	• Maintain continuity of private meeting space and mobile sales teams • Assess efficacy of mobile sales team efforts	• Terminate mobile sales teams (as needed) • Close temp meeting spaces • Return to steady-state sales operations
Marketing	• Alter marketing effort (website, social media, etc.) with new temp location and customer instructions	• Manage website and marketing effort under a displaced office • Develop marketing campaign for reoccupation	• Campaign to announce & celebrate reopening of damaged office	• Reestablish steady-state marketing operations
Policy and Compliance	• Create policy and compliance assessment of temp offices. • Adjustments as needed	• Develop plan for reoccupation of repaired office	• Policy and compliance adjustments as needed	• Reestablish and brief steady-state policy and compliance

Continued on facing page →

EVENT	Vacate Damaged Office Occupy Temp Offices	Recovery and Repairs of Damaged Office	Closure of Temp Office Space	Re-establish Original Office
Finance	• Create financial impact report for leadership • Reestablish finance security and compliance under temp office operations • Equip temp offices and mobile teams with expense accounts for operations	• Ensure timely payment of recovery contracts • Updates to leader-ship weekly on recovery expenses and cashflow • Plan for reoccupation • Report reoccu-pation expense to leadership	• Onsite for rapid accounts payables and receivables to manage cashflow through transition	• Reestablish steady-state finance operations
HR	• Survey employees and report displace-ment impact on our people • HR policy and support adjust-ments under temp location • HR support to administration for temp hires • Review and adjust insurance coverage for temp location and mobile sales teams	• Maintain pulse of employee engagement and work environment • Manage temp employee contracts • Develop plan for reoccupation of repaired office	• Changes to insurance and HR policies	• Reestablish steady-state HR operations

Synchronization may look and sound like the work of strategists with 50-pound brains who work on the 10th floor of tall office buildings, orchestrating complex planning sessions in support of large-scale operations. Believe me, it's not. You and your team synchronize each day as you tackle missions with little time, people, money, and other resources — where you place people and resources in places and in times where they may complement and support each other. Discipline is required to synchronize. Don't be blind to friction — those things that *could* happen that *might* derail your plan — but try to keep friction at bay as you synchronize people, teams, and resources. Success will be obtained in execution as you manage the day and the crisis, guided by planning, even if the plan is only sketched out on a legal pad. As we often said in the Army, "Even the best plan doesn't survive first contact with the enemy ... because the enemy gets a vote."

To close this chapter, I'll offer this perspective. Unless you work in a relatively large organization, it is unlikely you'll have enough people and resources to form a large, multi-faceted TOC, as we did in the military. It is also unlikely that you'll be able to perfectly execute all TOC functions as they are described in this chapter. That's okay. Excellent performance "left and right of the boom" happens in organizations of all sizes. The most important part is that you and your team understand the TOC functions and of the value of a TOC, especially under crisis. If your TOC consists of one person, such as an office manager, have that person read and demonstrate that they understand this chapter and describe how they will conduct TOC functions. Leverage the greater team (including you) to round out the individual you have designated as your TOC, requiring reporting, analysis, and decision making at a lower level to seize the initiative. Give your intent. It could sound something like this: "Team, because we are in a crisis moment, I'm going to establish a tactical operations center. Sam will lead this effort for us, supported by each of you. The purpose of establishing a TOC is to better manage this crisis and prevent further damage to our organization and hardship."

Short on people? Leverage technology. No organization has enough people to tackle its myriad tasks or to process and share information 24/7. Today, fortunately, there exists a surplus of technology to help, especially in a crisis. While technology may not think for us (though it may be headed that way), it can do wonders displaying the data we need to understand, visualize, describe, and direct. I have applications in my small business that allow me, its leader, to see where all my teams are in real time. My automated phone system allows me to take calls in bulk and read transcribed messages, so I can triage which customers to respond to when times get stressful. Other automated applications help me communicate via bulk email to my customers. I have bookkeepers and an accountant, but I also have automated systems that help me track cashflow — the very lifeblood of a business. Akin to a TOC, where a watch officer would sound an alarm, I can see receivables and payables on demand to know how resilient my business is and know when and where to shift funds to keep the business afloat.

Of course, there is nothing like a person with all their senses in full alert to help in times of crisis. Don't let the TOC concept overwhelm you. When crisis hits, figure out *who* can do *what*. Leverage technology, for sure, but engage in some vulnerability and ask for help from the team and from individuals for the benefit of all. Perhaps James can explore some back-up distributors to keep an important job on schedule. Maybe Mackenzie has some technology acumen and can troubleshoot an automation flaw or create a shared folder or alert for the team so that the business can weather the storm of a crisis. People will carry the day when crisis hits. Call on people even if you doubt that they have the maturity, intelligence, or even motivation to chip in. They will surprise you. Piece all this together — people (including yourself), technology, and other resources — to form your TOC. Ask for more but do place a timeline on it. You cannot operate in crisis mode forever. Get the team and your automated systems to step up and you'll overcome. The goal is to get "right of the boom" as quickly as possible, then busy with "left-of-the-boom" activities again. Ever onward!

There is nothing like a person with all their senses in full alert to help in times of crisis. Don't let the TOC concept overwhelm you. When crisis hits, figure out *who* can do *what*.

CHAPTER 8

Leading Change

Crisis and change seem to have family ties. Crisis has a way of highlighting the flaws of an organization — things that don't work as they should, or that cause eventual problems. In my view, crisis can and often does provide opportunity for growth ... and growth comes through change. Let crisis change you.

This is not absolute. Crisis does not automatically mean change (or at least not lasting or permanent change). There may have been some external influences that challenge your organization, such as a bad storm, that fortunately don't mean a wholesale revamp of how you run your organization. Some crises are a blip. Spend a moment contemplating this as a leader as you (and perhaps you alone) define and describe the environment.

Recall the "battle command" approach I mentioned in the previous chapter and apply it to determine the need for change in understanding, visualizing, describing, and directing. While the pandemic forced change for so many businesses, perhaps there were things that should have remained constant. For example, if a business succeeded due to its process for handling potential clients — managing them from initial call to paying client — it should not let the impacts of the pandemic change that. Pandemic safety requirements might have altered its engagement, shifting from in-person to virtual meetings to prevent viral spread, but the organization should stay true to what works. This is also true when results are not immediate. Many startup businesses operate at a loss for an initial term until they begin to turn a profit. Amazon operated at a loss in its early years, but its business model was sound — requiring the company and its people to stay the course, not altering its processes in hopes of an immediate remedy. (It helped that they had heaps of cash from investors to hold on during their ramp-up phase.) Before we dive into leading change, let's revisit Battle Command.

Restaurants are a great case study for our purpose here. Restaurants felt a massive impact when the pandemic hit. Accustomed to serving customers in busy dining rooms where a sneeze at one table can travel easily to another, the entire model of restaurants was ground to a halt by the initial shut-down of "non-essential" businesses and then the mandates for social distancing. There was no way a restaurant could get around not having people in its establishment ... at first. Things had to change. Here is an example of Battle Command for this scenario.

Understand. Define the problem. Our business, which requires customers to assemble inside of our dining room, is prevented from doing so. The problem statement (which should be in the form of a question) is, "How can our restaurant business deliver food and a relaxing and welcoming environment to customers if we are unable to seat them inside our restaurant?" I understand that this restriction may be in place for some time, and we do not have the cash on hand to remain closed beyond one week. I understand that my payroll is this amount and that I'll run out of cash to pay my workers in two weeks if we don't have new revenue coming in. I also understand that we have a proven demand — a loyal following of customers who will want to enjoy our menu selections and restaurant atmosphere, regardless of the pandemic.

Understand first. There are no viable solutions until you've understood the problem you're trying to solve.

There could and should be more to this, as a leader spends quality time — often alone or with a small group — understanding. Understand first. There are no viable solutions until you've understood the problem you're trying to solve.

Visualize. This is where an imagined future comes in. Knock down a predominant obstacle first. Remove the boundaries of your thinking. Reimagine delivering outstanding food, drink, and a welcoming restaurant experience to your customers. They receive what you deliver, just in a different way. Entrepreneurs do this. Who believed the Uber concept would have worked years ago — getting into a stranger's personally owned car and paying them through an app on your smartphone, without ever exchanging cash or handing someone a credit card? Remember the 5 Cs of Crisis Leadership, specifically confidence, from Chapter 2, and switch on your limiting beliefs radar. Visualize a scenario in which you solve the overwhelming problem, perhaps with innovative creativity.

Focusing first on the obstacles and getting caught in limiting beliefs is natural and the only way to overcome them is to, first, recognize them for what they are and, second, challenge them internally or with a team. For example, "We could shift our offering to delivery or external restaurant pick-up, but people will probably not elect to eat our food without the service they always rave about." Accepting this as truth does one thing: it stops all further exploration of possibilities. Why move forward from this point if you believe that customers will react the way you envision? Indeed, there is a dose of reality. If your imagined future requires moon landings and you don't own a spaceship, then you are probably a little off the mark. That said, I can imagine moon landings in the same way we land in airports all over the globe someday. Back to our restaurant and visualization.

Visualize. Close your eyes, then ask your team to close their eyes and visualize customers eating your delicious, expertly prepared food in the comfort of their own home or at a park, on a patio, or even in their car. Choose a favorite dish from your menu and imagine a customer enjoying it in a place of their choosing — perhaps on their couch watching a movie or on their porch enjoying the quiet and scenery. While the decorative elements of your restaurant and the unique dinnerware (your pretty plates or your awesome steak knives) are not in the picture, the enjoyment of every bite remains the same as if they were sitting at their favorite booth in the restaurant. Their appetizer is followed by a generous portion of their favorite main entrée while they dream of their favored dessert selection. Now describe how this happens.

Describe. In general terms (leave the creation and execution specifics to your team), describe — walking backward from the end state — how to satisfy customers in this pandemic "new normal" (or whatever the change) environment. "OK, team, a regular customer of ours just completed their meal — from appetizer to dessert and they are completely satisfied, just as they are when they dine in our restaurant. This time, they enjoyed their meal in the comfort of their own home. They understand that eating in our dining room, being served by our superb wait staff, is not an option right now, but they still desire our delicious food. Instead of being served on our unique dinnerware, they transported and consumed their meal in our to-go packaging, which kept their food warm and fresh. They experienced service akin to our dining-in experience only this time, through a convenient phone or app-based ordering system plus a pick-up or home-delivery process, which made them feel welcomed and special. Pick-up was well organized and convenient for the customer. We replicated our attentiveness when they called, helping them select items, suggesting wine pairing, which they might buy from a separate store to make their meal extra special. Their meal — eaten from a different fork, in a different location, picked up or delivered differently — still hits the bullseye. A satisfied and loyal customer remains!"

Note the imaginative and positive tone throughout this description. Missing (on purpose) are the "hows" of this experience. The customer experience is described free of obstacles (limiting beliefs). You, as a leader, just walked your team through a team win — a championship. Time to direct.

Direct. "OK, team, let's make that description happen!" Direct action. There are many challenges here but your team, properly led, will rise to the occasion. You could create teams, such as the kitchen, wait staff, and host or reception. Direct these teams to sketch out the steps required to deliver the experience you just described. Ask them to develop a list of resources needed to achieve this. Lastly, ask them to define the change that is required. For example, servers achieving customer experience through phone calls and parking-lot delivery instead of a dine-in service.

Apply this same methodology to your company. Swap "dining" with whatever your product or service is. Execute Battle Command yourself in relative peace and quiet but involve your team as you can. Once you've directed, lead change.

Leading Change: 8 Not-So-Simple Steps

It is widely known that people are resistant to change. I'm a non-conformist and while I consider myself very entrepreneurial and innovative, I, too, find I'm resistant to some change. It is a threat to my comfort. It causes me to have to work harder. I need to be led through change like everyone else. But change is easier if we've built a strong foundation and if we've done good work in the left-of-the-boom space before the crisis (as well as in the right-of-the-boom space after our previous crisis). Return to our crisis-ready foundation in Chapter 3 and recall how leadership is, at its core, simply influence. Nothing needs influence more than change, especially forced change due to a crisis. In my first book, *It's Personal, Not Personnel,* I introduce Dr. John P. Kotter's 8-Step Leading Change Model and how I used it when assuming the leadership role of a large organization on the cusp of significant change. I highly recommend Dr. Kotter's book, *Change,* as part of your right-of-the-boom preparation. Here, I'll introduce the 8 steps of this change model and offer my experience to help guide your leadership.[30]

Let's immerse ourselves in yet another fictitious scenario (though, as you'll see, one drawn largely from my own experience since purchasing a window-fashions company!). A company that sells and installs window fashions — shutters, blinds, and shades — experiences a string of miss-measures where products do not fit and thus must be reordered. The leader of the company and his managers discover root problems to be the fast pace at which the company is moving and a lack of knowledge and experience with a variety of products sold by the company. The leader can

---

30 John P. Kotter, *Leading Change,* Harvard Business Review Press, Boston, Massachusetts, 2012.

visualize a future not completely void of errors but where they are rare. He visualizes an organization moving at a steady (not hurried) pace, where teammates check each other and where they "leave no doubt" when it comes to getting orders correct. He describes this imagined future to his team, telling them they can rise to the challenge — to be exceptional. He then directs the team to implement steps in scheduling and ordering, which include extra oversight, group huddles, and customer interaction to decrease the margin of error. He leads change.

Step 1: Create a Sense of Urgency

Kotter tells us that the first step in leading change is to create a sense of urgency — to "Help others see the need for change through a bold, aspirational opportunity statement that communicates the importance of acting immediately." Urgency is not always the way. Some change can and should occur incrementally. In a crisis, however, a sense of urgency may be the leadership approach that saves the day. For leadership, urgency occurs in word and deed. Defining, announcing, and directing immediate bold action and seeing it through in execution is an example of this step. Kotter is describing a mindset here. Change must be not only recognized but felt by every member of the team.

Using our window-fashions company, here is an example of creating a sense of urgency. "Team, as you know, we have suffered a string of defeats when it comes to providing a flawless product and service to our customers. Failures such as these place the existence of our company at risk and none of us wants that. I take ownership for these failures and I'm asking for your help in fixing this. These changes must be implemented immediately. Immediate steps may not be perfect, but the company must act." Once directed, a leader can place themselves at the point of change — in this case, at that time and place where errors have occurred, such as where schedules are created and where orders are reviewed for accuracy.

Step 2: Build a Guiding Coalition

Kotter suggests that the second step in leading change should involve building a guiding coalition. He explains: "A volunteer army needs a coalition of effective people — born of its own ranks — to guide it, coordinate it, and communicate its activities." I'll call this assigning ownership or "pinning the rose" on individuals or teams to drive the change. Leaders must select those who are engaged with or psychosocially invested in the organization. These people or teams will become the champions of change and they must be directed as such and, moreover, embrace this role. "Alan, I want you to champion this part of our new method and report progress to the team. Can you do that for the team?" A coalition can be a single person, such as a deputy, or several people — the right people who will embrace the change. This is not a pass for leaders to unplug themselves from oversight, but it is an acknowledgment that leaders cannot drive change alone. They need a guiding coalition.

Step 3: Form a Strategic Vision and Initiatives

What comes next in Kotter's change model? Forming a strategic vision and initiatives. He tells us to "Clarify how the future will be different from the past and how you can make that future a reality through initiatives linked directly to the vision." Step 3 is outlined in the Visualize, Describe, and Direct concepts described in Chapter 7. It is important, especially under the direct portion of Battle Command, to include the "why" behind what is being directed. For example, "Team, our new scheduling process will bring in several minds, leveraging the strength of the entire team to help Emily schedule better [*why*]. We will review the following week's schedule as a group to help Emily manage our pace."

Step 4: Enlist a Volunteer Army

"Large-scale change can only occur when massive numbers of people rally around a common opportunity," Kotter tells us. Enlisting what he calls a "volunteer army" is the team approach to leading change.[31] Under the leadership of the Guiding Coalition — people wearing the proverbial roses — large groups of people can rally behind the change. It can be a form of pride-producing and healthy competition: Which team can demonstrate the change first? Be cautious of unhealthy competition, however, which may divide your organization. Celebrate those who rally

---

31 https://www.kotterinc.com/methodology/8-steps/

behind the change and figure out who is resisting so you can convince them the change is needed. Encourage those who are in support but have not yet produced results to "keep the faith." Insert yourself as a leader frequently during this change — in a variety of times and places — to see it through.

Step 5: Enable Action by Removing Barriers

Whether it's at work or on the highway, the only way to move forward (to "enable action") is to first remove barriers. Kotter tells us, "Removing barriers such as inefficient processes and hierarchies provides the freedom necessary to work across silos and generate real impact." Barriers come in several forms. When leading change, especially in crisis, removing barriers can be drastic. For example, giving a leader the authority to cancel all contracts under a certain dollar amount would be a drastic move. Canceling an existing internal procedure, protocol, or formal "policy" — like one that requires employees to seek the approval of a supervisor for specific activities — can be drastic. Removing barriers can mean authorizing expenditures beyond what has historically been allowed. A leader may remove a "no overtime" policy to provide more time for people to lead change. As Kotter states, sometimes a direct reporting chain that skips echelons of direct reports is needed to give the right focus and direction to something being changed. There are several examples. Removing barriers is often about simplifying the way you do business — flattening the hierarchy, eliminating committees and task forces, and reducing bureaucracy. What is important for leaders is that they identify what obstacles (such as restrictions, ineffective policies, or other barriers) stand in the way of change and to consider what resources may be needed to promote the freedom of action required to implement change.

Identify what obstacles (such as restrictions, ineffective policies, or other barriers) stand in the way of change and consider what resources may be needed to promote the freedom of action required to implement change.

Step 6: Generate Short-Term Wins

There is a reason why people who want to cross a finish line at a marathon first start with a 5K or a 10K. And if you had to wait all the way until your graduation to get the "summa cum laude" payoff for good grades, it would be hard to stay motivated for all those years — so colleges and universities (and high schools too) publish a "Dean's List" every semester to celebrate the short-term "wins" of their students. So too is it important to generate short-term wins at work, especially during change or crisis. Kotter tells us that "Wins are the molecules of results. They must be recognized, collected and communicated — early and often — to track progress and energize volunteers to persist." Short-term wins are essential, and, in some cases, leaders can even choreograph a win — they can toss a proverbial softball to a team member and then celebrate when it is caught. Leaders should identify where they believe success might be achieved, then plan to celebrate it with the greater organization when it does occur. In our window-fashions company example, the leader may highlight the transactions that went as desired and call them wins for the team. Praise and recognition are a big part of leveraging short-term wins. Praise even releases dopamine in the human body (that's why an "atta-girl" or "atta-boy" literally *feels* good), which fosters creativity in individuals. Keep your leader radar on for short-term wins — where you can create them and celebrate them.

Step 7: Sustain Acceleration

During change, you and your organization have places to go and no time to waste. It is not enough just to hit the ground running — you need to *keep* running and do your best to run *faster*. Kotter urges us to "Press harder after the first successes. Your increasing credibility can improve systems, structures, and policies. Be relentless with initiating change after change until the vision is a reality." I call this exploiting success. This does not mean remaining unsatisfied as successes are achieved — I'm not suggesting that your first successes are not "good enough." Indeed, it is important to espouse positive thinking and energy at every step; however, as leaders, we should always be thinking of how to *capitalize* on successes — how to solidify the change. Don't let the successes be isolated or fleeting, and don't let the team rest on their laurels once the first win is achieved. Celebrate it, then encourage your team to take the energy and motivation from a small win and apply it to the *next* one. In our window-fashions company example, the first flawless order and installation does not mean the team has achieved the desired change. *Several* flawless projects will indicate that it has. Sustain acceleration!

Don't let the successes be isolated or fleeting, and don't let the team rest on their laurels once the first win is achieved. Celebrate it, then encourage your team to take the energy and motivation from a small win and apply it to the *next* one.

Step 8: Institute Change

All these vital steps can still be for naught, in the long run, if you don't truly institute change — if you don't make it all indelible and ongoing. Kotter tells us that we must "Articulate the connections between the new behaviors and organizational success, making sure they continue until they become strong enough to replace old habits." I've always had a hard time placing an "-ed" at the end of change. Perhaps it is from my military experience, where we seemed to be in a period of perpetual change — where nothing was finish*ed* or complet*ed* and where we rarely had the luxury of saying we had succeed*ed*, won, or prevail*ed*. Making something final or past tense — attaching that "-ed" was difficult, if not impossible. This last step is an important one, though. I believe the best way to handle it — the best way to turn repeated successes into habits — is to approach them much like you approach small wins but, this time, you'll celebrate several small wins as one big one. You'll be able to demonstrate to your team that the change was worth it, that you are now on the other side of it, and, importantly, everyone must display vigilance to ensure the organization does not drift backward. Leaders can describe what things were like before the change and highlight the positive effects of change, congratulating all who had a hand in instituting the change. Be sure to describe the journey — these 8 steps, perhaps — to show everyone how it took place. This will help drive future changes, which are sure to come.

REFLECTION QUESTION

What in your organization requires change (which you know will meet resistance) and how can you lead it?

Ponder that for a moment. You don't have to move forward with it but the mental gymnastics you'll experience by imagining the steps above will be a good exercise for you. You may find that you can indeed implement and *lead* the change. Remember, people like comfort and order so you'll need to be convincing and persuasive. The methods offered here will help you. In the loneliness of your leadership position, ask and answer this question. Label the change, such as "requiring *installation technicians* to act as *salespeople* when in customer's homes" (that's one that meets resistance!) — asking them to share a sales pamphlet, describe the features of the product or service, gauge a person's interest, and attempt the sale. This initiative could be an effective way to mitigate the effects of a sales-slump crisis. Pick yours — your organizational change that is simultaneously required and unpopular among your people) and work through this model.

Is It a Crisis or a Change? Yes!

As you were reading about leading through change in this chapter, you might have wondered when to apply these concepts — during big changes or during crisis, or both? The answer is YES! Let me practice one of the techniques I learned in my tutelage as a crisis leader — repeating myself. As I stated at the opening of this chapter, crisis and change seem to have family ties — they are inextricably linked. Crisis will shine a spotlight on flaws, and it can and often does provide opportunity for growth, which occurs through change. Yes, crisis can be the perfect moment and motivator to put this chapter (indeed this book) into action. Outside of crisis, in times of order and relative peace, you and/or your team may discover things that may not work (a good left-of-the-boom practice). Change, in this case, is no less important or urgent. Outside-of-crisis change requires the Kotter approach just the same. You are in a storm or there is one coming, and it will be weathered far better with some institutional change. It behooves you, therefore, to remember that you have a choice of whether to be ready or not — of whether to embrace and lead change or to delay and resist it.[32] Spend a moment contemplating this as a leader as you (and perhaps you alone) define and describe the environment.

Crisis can and often does provide opportunity for growth ... and growth, as it turns out, only comes through change.

As you can see and as you surely know from your own leadership experiences, implementing and leading change is not automatic and it's not easy. It is incumbent upon you as a leader to determine if change is needed. Crisis is ugly and one way not to repeat it is to institute change. However, just because an organization experienced some ugliness does not mean systems and process are broken, the wrong team is assembled, or the company's vision is flawed. Sometimes bad things happen to rock-solid organizations; we learned that during the pandemic. Do the hard work to determine if, in fact, change is required. Hardening systems and/or people or strengthening an existing crisis-ready foundation may be in order but this may not require change on the scale of what Dr. Kotter subscribes. Assemble your trusted agents and do some analysis. If change is in order, follow the steps in this chapter. And keep the faith. Leading change takes courage, patience, and tenacity. Keep going!

---

32 For "the mother of all storm stories" and how they can create institutional crisis and spur long-term change, be sure to check out the story of JetBlue Airways' "Valentine's Day Massacre" in the book *High-Stakes Leadership in Turbulent Times* by Mike Barger, who authored the foreword to *this* book.

CHAPTER 9

Optimism and Opportunity

Even in crisis, opportunity lives and optimism is possible. I am an optimistic leader. That mindset has served me well in my Army and post-Army careers, even during my darkest moments. One does not need to look far in modern times for pessimism and negativity. Look at your phone, surf the internet, or watch the evening news if you feel a void. Crisis, indeed, is replete with negativity. No sane leader or organization wishes for it. Catchy sayings such as, "sometimes you win, sometimes you learn" or Albert Einstein's "A person who never made a mistake never tried anything new" don't carry much weight when sitting among the wreckage of a crisis. It's ugly, nasty, foul, chaotic, and downright painful, but optimism must survive if there is any hope for the future. Let me share some thoughts on optimism and opportunity and offer ways you can keep them close as you navigate your own crisis.

Even in crisis, opportunity lives and optimism is possible.

Choosing Optimism

Simon Sinek, famed speaker and author, calls himself an "optimistic leader." He even hosts a podcast entitled *A Bit of Optimism*. Sinek shared on social media: "Optimism is not being naïve or denying reality. It is a belief that the future will be better than our current situation." He contrasts positivity with optimism, claiming that positivity is stating that things are great when they are not (a tendency that can become toxic or harmful to others, diminishing the reality of difficult circumstances). Optimism is a belief. I can lean (and have leaned) on that belief to guide me as a leader. I frankly do not know how anyone can lead without optimism, for without it is to be without belief and hope — and organizations (and their people) need that belief and hope to get through the day, especially when those days might be ugly or overwhelming. I see optimism as more than just a personality trait. While I have it, I'm no different from any other person in that I am affected by the world around me and the reality of complex and difficult situations. My optimism is challenged daily, and it takes work to maintain it — real work. Optimism is a choice and a habit. It is a proverbial muscle we can strengthen by exercising it.

My optimism is challenged daily, and it takes work to maintain it — real work. Optimism is a choice and a habit. It is a proverbial muscle we can strengthen by exercising it.

My "optimism work" is more than just a "choose your attitude" proposition. The things I do in the left-of-the-boom space (and the people who I put at the center of it) create optimism in a truly strategic and ongoing way. To quote an African proverb: "To go fast, go alone. To go far, go together." The people I surround myself with fuel my optimism. I do the hard, painstaking work to find the right teammates. Indeed, I look for skills and experience to match the positions I must fill. That's the easy part. What I focus on mostly is presence and character. I'll ask questions in interviews to tease out a person's beliefs and see how they respond.

I spend time talking about the organization, its vision, and its values to see what reaction I get. I'll ask the individual at what point in their life they were most satisfied with the work they were doing. This helps me identify what they are truly passionate about. I'll ask what they *most* love doing and what they do *not* like doing. I'll have them imagine they are hired and ask them what they would want to be able to say about their contribution after a year with our team. I'll present dilemmas — such as when and why rules must be broken — and I'll ask them how they might approach situations like that. I'll present conflict and crisis-like scenarios to discover how they would handle such situations. I don't have "easy buttons." I've been burned by people who impressed me in the interview but who disappointed me down the road. The interview is just a start toward hiring and inspiring an optimistic team. I rely on my intuition, experience, and other teammates involved in the interview process to find people "like us" — those who we believe will espouse our values and commit to our organizational vision and, moreover, are optimistic. Our hiring process is followed by crisis-ready foundation activities to create engaged teammates.

My optimism is somewhat cheerleader-like. We don't hold Walmart-style pep rallies before each day, though I have been known to motivate my team by moving person to person, as if we were about to take the field during the playoffs (because we are), insisting that they give their all, have confidence in our vision and our values and in themselves. I can do this because I am genuinely motivated by what we do, and I love the people who do it — these circumstances are both foundational ingredients of optimism. My belief in our vision, mission, and people gives me the confidence and conviction that, when things go bad, tomorrow will be better. It is because of this that I can say with confidence, even in the darkest times, "Team, stay strong and focused, tomorrow will be better." It allows me to ask, "What are the opportunities?"

Opportunity Knocks

Just like optimism is a challenge in the darkness of crisis, so too is opportunity. But there *is* opportunity in crisis, and it behooves us not to waste it. Devoe's death provided opportunity. Indeed, I did not use that word in the moments following his death (and it makes me uncomfortable to use it here, knowing that many of you reading this book don't know me well and don't know my heart). I was certainly not thinking about opportunity as my people and I mourned his death, but I knew I had to honor his life and keep my unit focused on each other and our mission. Our opportunity following his death was to recommit ourselves — to steel our resolve, in his name — and to see the mission through and to bring our soldiers home alive. The Wharton School of Business at the University of Pennsylvania offers a useful and thought-provoking opinion piece written by Jerry (Yoram) Wind and Nitin Rakesh entitled, "Ten Guidelines for Creating Opportunity in a Time of Crisis."[33] It is a worthy collection of tools for your crisis kitbag. Here, I'll share seven of the 10 guidelines with my own thoughts, which I think are pertinent to our discussion about mastering of the art of leadership "left and right of the boom."

1. **Change your mental model.** A common, untrained, and unaltered mental model is, "This is bad and I want to bury and forget it." Change yourself and your team to a model of, "This situation, as ugly and painful as it is, presents an opportunity to learn, grow, and be the catalyst for enduring improvement."

2. **Step up your digital transformation and invest in infrastructure.** This refers to leveraging technology to overcome challenges, such as streaming live performances and offering digital menus to prevent the spread of diseases under a pandemic. "How would we digitize this?" would be an opportunistic question during a crisis.

3. **Identify new business opportunities generated by the crisis**. "Hmm, what if - hear me out people — what if this crisis is speaking to us in a strange, even spiritual way?" "What if we found some new and innovative business opportunities such as … ?" This is an example of how to approach this step — to spin the negative toward the positive and find the diamond in the rough. The shift from sit-down dining to take-out and/or delivery is a great example of an opportunity born of the pandemic.

---

33 Jerry (Yoram) Wind and Nitin Rakesh, "Ten guidelines for Creating Opportunity in a Time of Crisis," *Knowledge at Wharton*, The Wharton School at the University of Pennsylvania, April 14, 2020, https://knowledge.wharton.upenn.edu/article/ten-guidelines-create-opportunities-coronavirus-crisis/.

4. **Reexamine your talent strategy**. Do you have the right talent to weather reoccurring storms? Talent strategy means how you go about finding new talent and how you define talent — for instance: mature, educated, experienced, skilled, people of character. Of course, I tread lightly here. I see employees as people, not talent or human resources. This said, a business under crisis can do some creative thinking and imagining to then survive and thrive. Work-from-home, contractors versus full-time W2 employees, and even automated systems represent opportunities specific to the talent or people who drive the business.

5. **Identify areas of opportunity through an idealized design process**. Step way outside the box on this one. Maybe it is time to launch a new product line, bolt on an additional business to your existing one, or repurpose a position in your company by rewriting the job description and retraining an existing teammate. Let your greater vision, values, and culture guide you. Think wildly and boldly. You might find another diamond!

6. **Re-examine your business model and operations. Challenge your revenue model to identify opportunities for increased efficiency and profitable growth**. Indeed, people and organizational systems and processes will resist sweeping change but in the name of opportunity and optimism, crisis may just be the incentive to "go back to the drawing board." Maturity, prudence, and sound leadership are prerequisites as you search for opportunities. Dust off that original business plan. Ask yourself if it is still relevant — if the market analysis, cashflow model, etc. are still pertinent.

7. **Innovate and adopt an adaptive experimentation approach**. Assess risk, indeed — revisit Chapter 4 of this book — but experiment. Turn an existing process on its head. Develop something totally new. Sketch it out on a dry erase board first. Talk the experiment over with a diverse group of teammates. "Perhaps we can shift our hours." "Maybe we don't need a dining room after all." "What if we add this step in our customer experience process?" Take the opportunity to speed up the innovation process by involving more people in the process.

Good stuff, right? You might apply the ideas from Wind and Rakesh differently than I do, and that's OK. Your business is different from mine. But in the end, it's about having the right mindset, the right people, the right attitude about the pain and the purpose (and the opportunity!) of crisis, and being willing to reinvent your technology infrastructure, your financial models, and more — all in the name of delivering stellar products and services to your stakeholders.

REFLECTION QUESTION

What are/were the opportunity(-ies) in an existing or previous crisis?

I hope you are not currently sitting amongst the wreckage of a crisis. If you are, indeed, it's time to reflect and act upon this question and its answer. For everyone else, think back to a crisis you experienced in your personal or professional life. What opportunities did the crisis present formally or informally, consciously or subconsciously? What "diamonds" did you miss? Tease them out and use the Wind and Rakesh guidelines. This reflection and exercise will serve you well.

When the pandemic came on like a rogue wave in 2020, my leadership business — Rob Campbell Leadership — was showing momentum, especially on the keynote speaking and executive leadership coaching front. I had just returned from a retreat with my wife, Leslie, where I created the outline for what I intended to be my next book about the leader's role in the journey of an employee.[34] I was excited about the fast-approaching future. Then everything changed. Specific to my leadership practice, I could not visit offices anymore and conduct coaching in-person, as

34 While that book about the "leader's business" in the journey of the employee was intended to be Book #3 in the *Battlefield to the Boardroom Series*, the crisis of the pandemic drove me to set that budding manuscript aside to write this book, *Left and Right of the Boom*, for you first. Fear not — Book #4 is already underway as of the release of this book that you hold in your hands right now. That crisis was a detour and not a roadblock and, as you can see, it was full of opportunity.

I had become accustomed to doing. Keynote speaking froze in place as event after event was cancelled. What was more (and what couldn't be overcome with the whole "just deliver the presentation via Zoom!"), companies tightened their purse strings to deal with the uncertainty and, unfortunately, leadership development was on the chopping block. Overnight, no one was investing in the kind of work I offered. Of course, I believed the pandemic called for *more* (not less) of what I could deliver, but I faced a daunting challenge of convincing worried business leaders of that. It was time for me to practice what I preach about crisis. Time for optimism and opportunity!

Knowing these were essential and being the optimistic leader I was, I sprang to action. "What are the opportunities?" I asked myself and my team. We had long discussions and debates. I steered us away from limiting beliefs such as, "keynote speaking is dead" or "no company has a budget for leadership." Three opportunities emerged. First, I would shelve the book outline I had created for something more relevant. I had deep experience in crisis leadership and something to share. I was a battlefield leader and "crisis" was practically my middle name. The opportunity and possibility excited me. I immediately crafted the outline for this book and began my work. Second, my coaching engagements would shift to virtual (video conference) meetings instead of in-person coaching. This method would help the client (the leader) and me, as it was logistically easier for both of us. While not optimal, we could jump on a video feed and see each other and interact as coach and leader must do. I had experience as an Army Colonel and Brigade Commander later in my career leading via video teleconference, as I was forward in Afghanistan and part of my unit remained behind at Fort Campbell, Kentucky. Third, speaking and training could be done virtually. Even as venues closed and events were cancelled, organizations remained committed to assembling and they took to video conferencing to achieve their conference objectives. The business community became much more comfortable with virtual meetings, and I could leverage this shift by serving leaders and their organizations — anywhere at any time. It would be a steep mountain to climb. There were many downsides but dwelling on the negatives would not serve my team well. Revenue took a hit, for sure, but the opportunity was real. The crisis didn't require me to fundamentally shift my offerings. Leadership remained the predominant issue that my existing and new clients relied upon me for, pandemic or not. I was standing on solid ground. I just needed to shift my approaches. I exercised the Wind and Rakesh steps 1 through 3 highlighted above.

I've said it before, and I'll say it again: It's lonely at the top. You'll find yourself terrifyingly alone when crisis comes. I felt as if I were in a bubble when Devoe was killed. In that bubble, I wrestled with doubt and limiting beliefs. I had visions of failure. My confidence took a hit. And I was shocked and grieving. Through that natural experience, I knew I had to reach for optimism and opportunity. "We are going to get through this," I assured those around me. When I phoned my wife, Leslie, with the news and some advice on how to care for our families back home in Alaska where we were stationed, I vividly remember my first words. "We are going to be OK." In communication with my leaders, sharing my sorrow and offering my condolences, I reconfirmed my confidence in them, our mission, and our team. "Let Devoe's death steel our resolve to see this mission through and to bring our paratroopers home. You have my trust and confidence." Optimism and opportunity. Even in the depths of adversity.

You'll find yourself terrifyingly alone when crisis comes.

CHAPTER 10

Take the Offense!

"Defense never wins," a boss of mine said to me as we toured the battlefields of Normandy, France, in 2010. I pondered that argument as we walked the sacred ground, examining and assessing German and American battlefield actions and their consequences. The German defenses along the coast of France seemed impenetrable, especially from a force entering from the sea. Mines and wire obstacles were intertwined along the beaches and harbors. Clear machine-gun fields of fire gave the German defenders the advantage. Yet, the Americans and their allies broke through. We approached the beaches "on the offense" and while many lost their lives (9,000 allied soldiers in the first 24 hours), this moment is almost universally considered the successful "beginning of the end" of Adolf Hitler's horrific regime.

I'm writing this book in 2024, four years after the onset of the COVID-19 pandemic, which gives me additional reason to think about defensive measures versus offensive measures. During the first two years of the pandemic, amidst mankind's greatest defensive measures —face masks, social distancing, business shutdowns, hand washing and sanitation, and eventually vaccinations and other measures to prevent the spread and impact of an infectious disease — millions of people were infected, and millions perished. We were on the "defense" and "defense never wins."

Most crises — whether we're willing to admit it — are preventable. Of course, there are "black swan" events, such as the pandemic, which hit us totally by surprise. However, if the right people had imagined the unimaginable, more could have been done to prevent it or lessen its impact. Bill Gates talked about the specter of a global health crisis during the 2017 World Health Forum and suggested investing in vaccine development to enable the creation of "novel vaccines" in as little as a year, should we ever need to respond quickly to save millions of lives. He is largely credited with having warned the world about pandemics (and what they can do to supply chains, etc.), but not enough people listened, took proactive steps to prevent it or, better yet, went on the offense scanning the world for potentially dangerous and spreadable viruses, then containing and killing them where they begin or before they are born. We could have had a counterinsurgency against viruses — we could have gone "on the offensive" — but we did not. We never think the worst will happen to us, especially not on a massive, universal scale.

Indeed, crises are hard to predict and usually hit us by surprise. The response of organizations is typically defensive in nature. *Boom!* "Now what?" A business may throw up some barriers, such as firewalls defending against malware or extra safety gear to protect workers from on-the-job injuries. These measures are the defensive sandbags and body armor of a business, and they help, but they alone are insufficient. A modern motor vehicle has defensive measures installed in it, such as airbags, blind-spot detectors, and back-up cameras. While these defensive measures help, they're not enough in the face of bad drivers or bad road conditions. Skilled and attentive drivers, plus safe roads, combine to create an *offensive* approach, which is essential to preventing crisis. Defense alone is just not enough.

In Afghanistan, we knew that sitting on our fortified bases would not save our lives. We had to patrol around the immediate area on the offense, searching for and eliminating threats. We had to seek and destroy the enemy on our terms and in a time and place of our choosing. "Going on the offensive" is characterized by initiative, tenacity, and opportunity. It means, quite simply, getting off your duff — stepping forward and moving toward threats to find them, fix them, and finish them. For a business, taking the offense starts with *seeing*. "See yourself, see the battlefield, see the enemy," is a phrase used in the military specific to the "offense" and following the principles outlined in *The Art of War* by Chinese military general and philosopher, Sun Tzu. He stated, "If you know the enemy and know yourself, you need not fear the result of a hundred battles. If you know yourself but not the enemy, for every victory gained you will also suffer a defeat. If you know neither the enemy nor yourself, you will succumb in every battle."

"Going on the offensive" is characterized by initiative, tenacity, and opportunity. It means, quite simply, getting off your duff – stepping forward and moving toward threats to find them, fix them, and finish them.

For a private business, before it goes on the offensive, "seeing" starts with seeing itself. This means having a keen understanding of its purpose — its "why." Why does it exist? What makes it special? What is the cause of the business and its mission, which people can rally behind and preserve? This is important, indeed, for leadership, but also because the enemy (in this case, competitors), a fledgling economy, or market shifts can pull a business off course, away from its true purpose. Staying true to "see yourself" is good insurance in the offense.

In addition to seeing yourself, you must also see the battlefield. "See the battlefield" means understanding one's industry and market. Seeing the battlefield in military operations means bringing the battlefield into 3D. It is where a map comes to life. Commanders see relief — such as terrain features, avenues of approach (where foot soldiers, tanks and trucks can travel), electric grids, key infrastructure in urban areas, and even agriculture. They see these features in a spirit of taking the offense — gaining an advantage and seizing the initiative against threats or attacking them where they are weakest. For a business, it sees in 3D as well, only its terrain features and key infrastructure come in the form of industry data, product performance data, consumer-demand indicators, and the like. A hotel aiming to take the offense would first examine and study economic data. It would also want to "see" its surrounding area, what it is conveniently located nearby, tourist information, traffic patterns (especially during holidays), and what informs the decisions of people when booking hotels. A restaurant would take the offense by first "seeing" itself but by also seeing its battlefield — convenience to drive-by traffic, proximity to shopping and tourism, and proximity to competitors. These are but a few of a large array of "battlefield" visuals any business leader must consider. Take to the offense by seeing and understanding first.

REFLECTION QUESTION

Where are your blind spots?

One of the most common phrases uttered in the aftermath of a crisis is "We didn't see it coming." But predicting and preparing for crisis is all about seeing — seeing yourself, seeing the battlefield, and seeing the enemy. Take a moment (and consider sharing this reflection with teammates at work) to ask yourself, "What are our blind spots? What are we not seeing clearly when it comes to our own strengths and weaknesses? What do we refuse to see and acknowledge about the marketplace and the competitive arena? And what don't we have insight into as it regards our competitors or other threats/enemies?" Making a list of things that "Well, clearly, we can see that ..." and get busy figuring out how to address what's right in front of you. See it, then slay it.

"See the enemy." In my civilian businesses, I've always been one who focuses on my vision and my "why" instead of focusing too heavily on my competition. I personally believe it's unhealthy to look at competition as the enemy. I don't feel I'm being naïve in this regard; I just consider it a better philosophy or at least one that better suits my leadership style. I certainly don't enjoy being seen as the enemy. I believe — on my small business battlefield, home improvement — that there are plenty of customers for us all. Home Depot and Lowes are neighbors and competitors in my city, and both seem to be doing just fine. All this said, I *do* want the customers my competition takes from me or gains before I do. I want to win the customer more often and earn a larger "share of wallet" for each customer. Moreover, I want to be resilient and ready for a crisis. Ignoring competition in a private business can be suicidal, therefore it is important to "see" them. What do they do differently than you? What would be *their* SWOT assessment? Where are they strong or weak? What opportunities might they seize when we least expect it and what threats might they be vulnerable to (that we can exploit)? Why would a consumer choose them over you? Apply the "see yourself" methodology as best you can when "seeing" your competition, because — in the end — the only way to truly see yourself or your competition is to see both of you in the context of the other. Understanding and acknowledging everyone who's on the proverbial battlefield will help you navigate the offense in a more informed manner.

Apply the "see yourself" methodology as best you can when "seeing" your competition, because — in the end — the only way to truly see yourself or your competition is to see both of you in the context of the other.

Let's look at an example of this in action. ABC Home Lighting might focus on higher-end homes and serve people of higher income brackets while its competitor XYZ Home Lighting prides itself on budget-friendly options. ABC, knowing and remaining true to its "why" yet wanting to gain more market share, could go on the offense. It could examine (as best it could) XYZ's target customer, defining them in as many categories (gender, age, income, attitudes, etc.) as possible. ABC could study the advertising and website of XYZ to better understand their message and market segment (lower-income or bargain-hunting consumers). ABC, to build resiliency and ready itself for a potential economic crisis, may offer but not advertise or focus on budget lighting options — an offensive move.

No Napping in the Trenches

Staying in business — in any type of organization and in any industry — sometimes comes down to never getting lazy or comfortable, and never thinking that protecting your business is a game of "defense." Follow the philosophies of Sun Tzu and you'll be ready for the *offense*. Think of your business offense in the same way I thought of mine. Don't sit idle inside the sandbagged walls of your base (your business). Get up, get out, and seize the initiative against the next crisis. Getting up and out means dedicating time — away from the madness of the business, reexamining SWOT, "seeing" your competition, and hardening your business in an offensive way. It means getting out to network and talking to customers in the field (where the point of sale occurs) to pick up on desires and changes in consumer behavior. Being on the offensive requires that you're willing to spend on the offensive too — that you don't under-invest in the future of your business, only spending money when you "must" (i.e., only spending money on the basics or on defensive measures).

Don't sit idle inside the sandbagged walls of your base (your business). Get up, get out, and seize the initiative against the next crisis.

There are so many ways to stay on top of things in your business — to stay awake in the trenches, planning your next offensive move. Read the local paper and watch the local and national news to never lose sight of the macroenvironmental and macroeconomic factors that impact your business. Stay engaged with local and even national government happenings so you can "see" laws and policies long before they become consequential to your organization. Leverage your local Chamber of Commerce to build business resiliency. Engage in guerrilla marketing — unconventional ways to gain more awareness and market share. Tell your people, "We are not in the [fill in the blank] business; we are in the marketing business." Get them active in talking about your product or service. Get them active looking for threats. Then act upon those threats by playing out fictitious scenarios, including how your business, its systems, and its members would react and how the threat may counteract. Once you've mastered the fictitious scenarios, get out there in the marketplace to tackle the real threats. This is the *offense.*

Taking the offensive can be risky. It may cost money. Thrusting yourself into places where threats emanate or where ugly data may come to light — such as engaging with the customers of your competition or attending government meetings open to the public — might backfire or bring bad news. But it is here where you can learn, seize the initiative, and, moreover, take action to protect against the next crisis. Taking the offense does not always involve visiting the battlefield. In the safety of your own office, you can run threat scenarios, and put your imagination and the imagination of your greater team to work imagining the unimaginable. But the battlefield awaits you. Lest you forget.

Conclusion

Throughout this book, I've asked you to engage in "reflection questions" that I think will support your growth, harden you, and better prepare you and your teams to prevent, mitigate, and respond to crises of all kinds. And now, I'm going to invite you to join me in ruminating on a question I've been reflecting on for more than 15 years.

Following the death of Patrick Devoe, was my unit in a state of crisis?

As a student of leadership, I wrestled with the answer to this question for years as I reflected on this crucible in my life — this devastating moment that forever changed my life and the lives of so many others under my command. I wrestled with it in the creation of this book. At face value, the answer is a resounding *yes*. We experienced a fatality, which clouted every member of my organization; it was the heaviest of all blows. The answer, however, is a resounding *no* if one examines *how* it impacted our mission. That IED explosion did not take us off course or cause us to make sweeping changes to our mission or methodologies. We were well-led across the board, and our mission and vision were stronger than anything the enemy (or any individual crisis) could throw at us.

I keep coming back to definitions. On the side of *yes* is the *Oxford Languages* definition of crisis — a *time of intense difficulty, trouble, or danger*. Devoe's death was indeed a period of intense difficulty, trouble, and danger for my organization. It hurt deeply and shockingly. It colored our dreams (or nightmares). We were fresh in a new combat zone, anxious, energetic, and relatively inexperienced. Losing Devoe was a reminder of the life-and-death nature of the very mission we undertook. On the side of *no* (no, this was not a crisis) is the *Merriam-Webster* definition — *an unstable or crucial time or state of affairs in which a decisive change is impending* ... and *a situation that has reached a critical phase.* Indeed, we experienced momentary instability while reacting to and recovering from Devoe's death. It was surely a crucial time for my organization and its leaders, but it would not result in "decisive change" or even a critical phase for my organization. The mission continued, undeterred. We were bruised but not broken. We had lost one of our own, but we had not lost our way.

So, was this particular "*Boom!*" a crisis? Yes ... and no. Patrick Devoe's death was a significant, tragic setback for the organization — for the 1st Squadron, 40th Cavalry Regiment of the United States Army — which brought momentary pause and threatened the confidence and focus of our unit, its mission, and its people. I didn't offer that exact definition to my unit, but I did indeed — formally and verbally — recognize Devoe's death as a setback and a threat to our confidence. I countered by caring for my soldiers and leaders and — with my wife, Leslie's, help — their families back home. I countered by promoting our just cause, our mission, our vision, and our people.

Grappling with the crisis question may seem like over-examination but it is not. It is precisely what leaders do. Definitions can be remarkably effective in simplifying things for people. And, in times of crisis, clarity and simplification are paramount. A leader may be the only person in their organization to examine, in greater depth, the condition of their organization, the environment, and how they may alter their leadership. A leader may be the only person in their organization to consider and implement the concepts shared in this book — to get left and right of the boom. It's lonely and important business.

The death of Patrick Devoe, a significant and lasting tragedy, would not signal a crisis for my unit so long as leadership did its thing. It could have been a tipping point if leaders did not move quickly to assuage the grief of their soldiers and keep everyone's focus on the mission. The risk was high. I had to act and communicate decisively, and I did. We were *new* to Afghanistan, as you know. Most of my soldiers and many of my leaders had never been in combat before. We were a tight team (a family, truly) and one of our own had just been taken. Confidence was at risk — something you will witness in your own crises, I assure you. Void of proper leadership, soldiers and their teams could have become overly cautious, taking their eyes off our counterinsurgency objectives (which meant thrusting ourselves into harm's way). Vengeance, a normal human reaction to having an enemy hurt or kill someone you care about, could have seeped into the ranks, causing soldiers and leaders to overreact. But it did not, and we did not. The immediate demands of our transition with an outgoing unit and our critical overall mission would not allow the proper time to grieve our loss. It wasn't fair and it wasn't easy. The transition under way would require our full attention and focus. It would require us to remain at the top of our game. Additional fatalities, perhaps several more of them, could represent a tipping point or a significant situation requiring radical changes. I had seen that in combat, but my unit was not there yet. *Loss* had happened; *crisis* was still being held at bay.

I believe that what was true for my unit can be true for your organization too — that tragedy may not turn into crisis so long as leadership does its thing.

How We Keep Going, in a World Full of Crises

Using Devoe's name and death as the catalyst for this book and labeling his death a setback has weighed heavily on me. I have wondered, "What if one of Patrick Devoe's family members reads this book? Or what if one of my other soldiers, who knew and loved him, takes issue with how I have recounted that awful day and what preceded and followed it? How will they feel and what will they think of me when they hear me refer to Devoe's death as a 'setback?' How will they feel at the realization that I did not focus on our other fallen brother in arms, Nick Gideon, honoring *his* life and telling *his* story?

The names of our military fallen are forever enshrined on gravestones and monuments across our nation and the world. There is something more, however, to these monuments of granite, steel, and stone. Something spiritual, I believe. Something living and enduring. It is the voice — the *calling* — of the fallen to us, the living. It is captured in the poignant war poem "In Flanders Fields," by Lieutenant Colonel John McCrae, which commemorates the military cemeteries of Europe. "The torch; be yours to hold it high. If ye break faith with us who die, we shall not sleep ..." The fallen walk and talk no more yet their voices live. They have something to promote and to teach us — patriotism, personal character, leadership, and citizenship — a *United* States. I know enough of Patrick Devoe to say that he would want us to live our lives and to care for our people and our organizations as he did for his squad. This book is his torch, now carried by me, handed to you in the form of the lessons captured in this book.

If this book was a *memoir*, I would not have included leadership insights or practical tools for preparing for and responding to crisis. And I would have shared a great deal more with you about the gut-wrenching experience of losing a soldier when you're their commander — about the emotions I felt and the way I coped, week to week and month to month, in the aftermath of Devoe's death (and of Nick Gideon's death in July of 2009). Again, I defer sympathy toward those who knew and loved him best. Instead, I offer the business book, *Left and Right of the Boom*, as a torch, written for and to be carried by business leaders who need to know how to get above the fray and lead, even in the worst of circumstances. One way of keeping faith with Devoe — to honor his life — is to focus on the realities of how *organizations* live and die — how organizations get offensive and defensive, how they become creative and calculating, and how they get sharper in the lead-up to crisis and wiser in the aftermath ... all in a virtuous cycle of leading their people and operating their organizations in good times and in bad.

For all these reasons and more, I have come to peace with using words like "setback" when something bad happens in an organization and the organization is able to move forward without requiring major changes — when what feels like a "crisis" in the moment doesn't take the organization down. But thinking about painful moments in our organizational histories as something other than a "crisis" if it doesn't quite fit the definition can take time and practice. For me, when it came to the loss of Devoe, it took rational thought (not pride and ego), guided by definitions, and it took a deep understanding of my unit and its people (and the environment we were in) to land on "no" to this important question. Not every boom is a crisis, but every boom hurts and every boom requires that we do the work to understand it, learn from it, and not repeat it.

In 2009 in Afghanistan, we got left and right of the boom. We practiced what I have shared in this book. We established a rock-solid — yet flexible — foundation, which would carry us through any storm. We invested greatly in our vision, mission, and values, plus our people and their families. We disciplined ourselves under an intense and demanding schedule to ready ourselves for a crisis. When the boom happened — as it did with Devoe and several other occurrences in a year in combat — we practiced the 5 Cs of Crisis Leadership. We exercised battlefield patience, and this reasoned, mature, and composed approach permeated our organization. We asked ourselves hard questions, like "How do we define this event?" and "Will it lead to our eventual demise?" We formed formal and informal crisis-management teams (CMTs) to examine challenges and guide us. We created risk assessments — ad nauseum — and these exercises saved lives yet allowed us to continue our missions. We got countless things wrong. Indeed, this book is a product of my failings and the failings of my unit; but following these failings, we conducted formal and informal after-action reviews (AARs) to capture lessons, learn, and grow. In the end, we got many things right.

Under the Afghanistan sun and moon, while mourning and soldiering on, we communicated formally and informally through several means and with several stakeholders using SITREPS (situation reports) or elements of them. Our communication increased and strengthened when under crisis. We were human beings, not superheroes, and our confidence needed cultivation; each of us, in private moments, wrestled with it. I and my leaders took note when confidence was shaken in individuals and in teams, and we went to work repairing and boosting it. We had formal tactical operations centers (TOCs) but in moments where we did not (in training events and back at home station in garrison), we followed the concepts of a TOC in all that we did — exercising curiosity, striving for synchronization, and receiving and disseminating information. We led change when change was called for and we did yeoman's work to find and choose optimism even in our darkest moments. We defended ourselves for sure, but we never settled in the defense. We took the fight to the enemy or went on the hunt for threats, whatever their form, to fix and finish them before they arrived at our doorstep.

We did all this at scale, formally and informally, in small teams of three or a task force of more than 700. So, too, can you apply these "left-and-right-of-the-boom" tactics to your teams, task forces, committees, and organizations, regardless of their size. I personally took these concepts into entrepreneurship, and I practice them today in my small businesses (one that comprises five people and the other that employs just four). It is my hope that you will take these lessons from the crucible of crisis — the battlefield — and incorporate them in your organization. Make them your own but make them sacred. Remember, while you will likely not follow every step outlined in this book in expert fashion, you'll be guided by these concepts, nonetheless. And that's a great place to start. I'm betting that these stories and concepts will speak to you, even if only subconsciously, when the time comes (*Boom!*) and they will help you and your organization survive and even thrive.

There is perhaps no higher calling than military service — no more noble a profession than the profession of arms. I was privileged to have served in the ranks of my nation's Army — to have led men like Patrick Devoe. We servicemembers swear an oath do a document, the United States Constitution. We surrender our individuality for that of the greater team and the mission. We write a blank check payable to our nation with our lives. The United States of America cashed Patrick Devoe's check. His death was tragic and any of us in 1-40 Cavalry Airborne (indeed, his family) would give anything to have him back — to not have to tell this story. We would give anything to have Gideon back.

What are we to do when lives are lost? I don't have a guidebook for that or a way to make it any less painful. What I do know is that we honor those lost by living good lives. We take the lessons they left for us; we listen to their enduring voices and learn and heed the lessons they have left us. We carry the torch so that they may sleep. Among others, Patrick Devoe's lesson is one of love and leadership, one of teamwork and comradery, one of being ready for crisis.

You cannot see, smell, or feel the literal blood, sweat, and tears behind the lessons captured in this book — my three decades of leading and learning on battlefields and in boardrooms — but this book is soaked with them. I have been a soldier and a civilian, and I have loved both. Whoever you are and wherever you lead, I honor the difficult and important work you do, and I hope you love it as much as I have loved my own careers. I hope the crises you have weathered have strengthened you instead of breaking you, and I hope this book helps you fare better next time.

I think of Devoe daily as I brew my coffee. I carry the lessons he taught me into private business. I honor his life with this book.

Those I lead today never knew Patrick Devoe, but they are the recipients of the lessons he taught me. They operate — with me as their civilian "commander," ever at their side — left, right, or in the boom when it happens. And no, running a window-fashions company and a residential painting company is *not* a whole lot like running a combat task force. There are no body bags. There is something deeply serious when lives are at stake, as they were every day when I was leading combat troops. The life-and-death risk is something that captures our attention, energy, and affection like nothing else. Those of you who work in healthcare or high-risk industries like aviation or nuclear energy may understand this better than most. Thankfully, for most of you, your businesses don't operate in places where you might hit an IED. But that's no reason to be complacent about crisis. Your next workplace crisis may not result in injury or death, but its effects and impact will be no less serious on the organization — no less serious than the injury or loss of a soldier to an IED. Crisis can and will shake you and your business just the same as it did my unit on the morning of March 8, 2009. The question remains, then: How well will you lead and manage it? Get left and right of the boom, my friend ... because when the boom — no matter how you define it for your team — comes (as it surely will), you'll want to meet it, defeat it, learn from it, and persevere. And you can. Make no mistake about it: Your next crisis is coming. Now strap on that body armor. Here is your torch. Go lead!

Acknowledgments

I do nothing alone. Nothing. This book is, indeed, decades in the making, with so many people to thank.

I thank **the soldiers** — superior and subordinate alike — of the United States Army who underwrote my mistakes, took a chance on me, taught me, and allowed me to lead.

Special thanks to **the paratroopers of 1-40 Cavalry**. We traveled a journey together of laughter, loss, learning, and love, which I will cherish for the rest of my days.

My loving and supportive family, **Leslie, Robbie, and Louden,** who are my strength and inspiration.

The team behind the creation of this book — **Kate Colbert**, my editor and publisher with Silver Tree Publishing and teammate since the birth of the *Battlefield to Boardroom Series*, and **George Stevens** with G Sharp Design for his remarkable book cover and interior design. **Samm Sawyer** with BizCo Studios for my branding. I'm proud of the work we did together to make this book the best it could be.

Thanks to **Mike Barger**, a fellow military veteran and business crisis leader who readily accepted and wrote a superb foreword — a fitting opening to this book.

I thank **my teammates in All About Blinds, All About Painting, and Rob Campbell Leadership,** who continue to teach me and who rise to the occasion when things go "boom!"

Keep In Touch with Rob!

- **RobCampbellLeadership.com** – An evolving and relevant site where you'll find videos of Rob speaking about leadership and veteran's issues. Or visit his blog page for frequent, useful "battlefield to boardroom" content. Here you can find his books or just stay in touch!

- **LinkedIn.com/in/RobCampbellLeadership** – Rob is there daily ... posting, responding, and sharing useful content on leadership, crisis management, veteran and military spouse transition issues, and entrepreneurship. Please connect!

- **Facebook.com/RobCampbellLeadership** – Follow Rob, send him a message, and join the conversation.

Go Beyond the Book

Hire Rob to:

- **Speak and Inspire.** Rob speaks and inspires audiences on a variety of leadership and veteran and spouse advocacy topics. Contact the HeyRed Speakers Agency at https://heyred.biz/.
- **Deliver Impactful Leadership Training.** Rob can train your leaders on contemporary leadership approaches for today's volatile, complex, and increasingly virtual business environment.
- **Coach You or Your Colleagues to Success.** Rob coaches leaders and veterans on their leadership or transition journeys.
- **Prepare You for the Realities of Crisis Management and Leadership.** Rob can help you understand how to lead before, during, and after crisis. With deep expertise in leadership and management under chaos and drastic change, Rob can bring strategy, calm, and compassion to organizations turned upside down.

Get the conversation started at Rob@RobCampbellLeadership.com.

About the Author

Colonel Rob Campbell (USA, Retired) entered the Army in 1990 and served a 27-year career as an infantry officer, ranger, and paratrooper, commanding infantry and cavalry units from platoon to brigade. Rob is a veteran of the wars in Iraq and Afghanistan, with service in the 25th Infantry Division and the 82nd and 101st Airborne Divisions. He also served during Hurricane Katrina relief efforts in New Orleans, Louisiana. Rob's core purpose in life is to make a difference in the lives of others through optimistic leadership. Clear on this purpose, Rob decided to retire from the Army in 2016 and continue to give back to the community at large.

Rob wrote his first book, *It's Personal, Not Personnel: Leadership Lessons for the Battlefield and the Boardroom*, in 2017 and became an entrepreneur, founding a leadership coaching, speaking, and consulting business.

Rob's second book, *At Ease: Enjoying the Freedom You Fought For — A Soldier's Story and Perspectives on the Journey to an Encore Life and Career,* was published in 2020. Rob is a staunch supporter of veterans and their spouses. He is the Chapter President of The Veteran Business Collective, Tri-Cities, Tennessee. Rob leads an effort called Salutes and Handshakes, an initiative to attract and retain talent in Eastern Tennessee.

Today, along with his leadership practice and publishing company, Rob owns and runs two small businesses in Johnson City, Tennessee. All About Blinds, LLC, is a window-fashion company that sells shutters, blinds, shades, drapery, closet systems, and awnings. All About Painting is an all-surface interior and exterior, residential and commercial painting company. Both companies serve the greater Eastern Tennessee region, including North Carolina and Virginia.

Rob shares his lessons — from the battlefield to the boardroom — on stage in front of businesses and government organizations, coaching business executives and training leaders. He is a classic car (1966 Mustang and 1986 Bronco) collector and enjoys traveling, reading, and writing.

Rob is a native of Massachusetts and graduated from the Massachusetts College of Liberal Arts in North Adams, Massachusetts. He has Master's degrees from Central Michigan University and the United States Army War College. Rob is married to the former Leslie Ann Hamp. He and Leslie have two sons, Robbie and Louden. Rob and his wife, Leslie, reside in In Flag Pond, Tennessee. *Left and Right of the Boom* is Rob Campbell's third book.

Loved *Left and Right of the Boom*? Don't miss the rest of the book series!

Before there was *Left and Right of the Boom*, there was *At Ease: Enjoying the Freedom You Fought For — A Soldier's Story and Perspectives on the Journey to an Encore Life and Career*. Heralded by military leaders as **"A practical, inspirational, fun, must-read book"** and **"Groundbreaking literature for the post-9/11 generation, the 'Learning to Eat Soup with a Knife' kind of guide to life after the military,"** *At Ease* offers transitioning service members candid and practical advice about everything from incorporating your family into your transition to celebrating your service before putting it behind you. **Available in paperback and ebook editions.**

Rob Campbell's debut book, *It's Personal, Not Personnel: Leadership Lessons for the Battlefield and the Boardroom*, introduced the world to the Colonel's proven approach to leadership. It's more important than ever — in corporate America and in the American military — for leaders to understand how to invest in people. In his first book, for a broad audience of leaders in every environment, Rob Campbell offers true stories and practical frameworks you can easily apply. By understanding that leadership is personal, you will discover that all workplace challenges — productivity issues, efficiency issues, turnover problems, lack of employee engagement — come down to whether employees are being acknowledged, valued, and understood. **Available in paperback, ebook, and audiobook editions.**

Made in United States
Cleveland, OH
04 November 2026

47644439R00126